CHICAGO PUBLIC LIBRARY
BUSINESS / SCIENCE / TECHNOLOGY
400 S. STATE ST. 60605

R00171 93482

CHICAGO PUBLIC LIBRARY
HAROLD WASHINGTON LIBRARY CENTER

R0017193482

THE COMMONWEALTH AND INTERNATIONAL LIBRARY
Joint Chairmen of the Honorary Editorial Advisory Board
SIR ROBERT ROBINSON, O.M., F.R.S., LONDON
DEAN ATHELSTAN SPILHAUS, MINNESOTA

SELECTED READINGS IN PHYSICS
General Editor: DR. D. TER HAAR

NINETEENTH-CENTURY AETHER THEORIES

TO GABRIELLE

NINETEENTH-CENTURY AETHER THEORIES

by Kenneth F. Schaffner

Associate Professor of Philosophy and of History and Philosophy of Science,
New Collegiate Division, The University of Chicago

PERGAMON PRESS
OXFORD · NEW YORK
TORONTO · SYDNEY · BRAUNSCHWEIG

Pergamon Press Ltd., Headington Hill Hall, Oxford
Pergamon Press Inc., Maxwell House, Fairview Park, Elmsford,
New York 10523
Pergamon of Canada Ltd., 207 Queen's Quay West, Toronto 1
Pergamon Press (Aust.) Pty. Ltd., 19a Boundary Street,
Rushcutters Bay, N.S.W. 2011, Australia
Vieweg & Sohn GmbH, Burgplatz 1, Braunschweig

Copyright © 1972 Pergamon Press Ltd.

All Rights Reserved. No part of this publication may be reproduced, stored in a retrieval system, or transmitted, in any form or by any means, electronic, mechanical, photocopying, recording or otherwise, without the prior permission of Pergamon Press Ltd.

First edition 1972

Library of Congress Catalog Card No. 77-133397

Printed in Hungary

This book is sold subject to the condition
that it shall not, by way of trade, be lent,
resold, hired out, or otherwise disposed
of without the publisher's consent,
in any form of binding or cover
other than that in which
it is published.

08 015673 8 (flexicover)
08 015674 6 (hard cover)

CONTENTS

PREFACE vii

PART 1

I. Introduction: the Functions of the Aether 3

II. The Historical Background of the Nineteenth-century Aether through Young and Fresnel 7

III. Aberration from Bradley to Michelson and Morley 20

IV. The Elastic Solid Aether 40

V. The Electromagnetic Aether 76

VI. Lorentz' Aether and the Electron Theory: the Electrodynamics of Moving Bodies 99

REFERENCES 118

PART 2

1. Letter from Augustin Fresnel to Francois Arago, On the Influence of the Movement of the Earth on Some Phenomena of Optics 125

2. On the Aberration of Light, by G. G. STOKES 136

3. On the Relative Motion of the Earth and the Luminiferous Aether, by A. A. MICHELSON and E. W. MORLEY 144

4. On the Laws of the Reflexion and Refraction of Light at the Common Surface of Two Non-crystallized Media, by G. GREEN 161

5. An Essay Towards a Dynamical Theory of Crystalline Reflexion and Refraction (Sections I and III), by J. MACCULLAGH 187

6. On a Gyrostatic Adynamic Constitution for 'Ether', by W. THOMSON (Lord Kelvin) 194

7. On the Electromagnetic Theory of the Reflection and Refraction of Light (an extract), by G. F. FITZGERALD .. 204

8. The Rotational Ether in its Application to Electromagnetism, by O. HEAVISIDE 208

9. *Aether and Matter* (sections 49–54, 102–16), by J. LARMOR 213

10. An Inquiry into Electrical and Optical Phenomena in Moving Bodies (Introduction), by H. A. LORENTZ 247

11. Simplified Theory of Electrical and Optical Phenomena in Moving Systems, by H. A. LORENTZ 255

INDEX 274

PREFACE

THIS book is an account of a group of theories which occupied the attention of some of the best physicists of the nineteenth century, but which is largely only of historical and philosophical interest today. During the previous century the various aether theories held an importance for physicists which general relativity and SU 3 symmetry, for example, hold today. Fresnel, Cauchy, Green, MacCullagh, Stokes, Kirchhoff, and Lord Kelvin, to mention only a few physicists, developed elastic solid theories of the aether. Maxwell, Fitzgerald, Heaviside, Sommerfeld, and Larmor engaged in serious research on the mechanical characterization of the electromagnetic aether. The famous Michelson and Morley experiment, when first performed, was thought to support Stokes' theory of aberration rather than Fresnel's view. Subsequently the experiment was reinterpreted to support the Fresnel–Lorentz theory of the aether. It took over two decades before the interferometer experiment came, after Einstein's work, to signify a confirmation of his theory of relativity—a theory which employed no aether at all.

The analyses in this book constitute the first stage of an investigation into some of the important ideas and experiments which led to Einstein's special theory of relativity. Certainly, though, the analyses of the aether theories and the various experiments associated with them can be considered on their own terms, rather than as leading toward any particular goal. For reasons which I shall spell out in some detail in the first chapter, I believe that this book should be of interest to scientists, historians of science, and philosophers of science.

Selections from among what I believe to be the most important primary sources in aether theory constitute the second part of this book, and of course they are necessarily selective. Sufficient references should be given in the commentary of Part 1, however, to guide the reader to most other primary sources. There are omissions of some aether theories, such as the vortex sponge aether and C. Bjerknes' and A. Korn's pulsating sphere theory, and the interested reader is advised to consult E. T. Whittaker's (1960) important monograph on the history of aether theories for a more comprehensive account. Though Whittaker's work is the most important secondary source on the history of aether theory, and though this author is in great debt to Whittaker's book, I cannot advise that it be read alone. There are some severe defects in Whittaker's monograph, particularly in connection with his unwarranted idolization of MacCullagh's aether, his failure to consider many developments in aether theories during the years 1860–1880 in both England and on the Continent, and—in what is perhaps his most famous blunder—his unfair treatment of Einstein's special theory of relativity. For these and other reasons I have found that Whittaker had to be supplemented with other secondary works to obtain an accurate view of nineteenth-century aether theories. The most useful of these supplementary works are in the *Reports to the British Association for the Advancement of Science* by H. Lloyd (1834), G. G. Stokes (1862), and R. T. Glazebrook (1885). A book by H. A. Lorentz (1901) which re-presents a series of lectures which he delivered in 1901–2 on various aether theories and aether models is also very useful. Essays by Rosenfeld (1956) and Bromberg (1968) have also proved to be stimulating and helpful in obtaining a more accurate overview of developments in the electromagnetic theories. Papers by G. Holton (1960, 1964) and by T. Hirosige (1962, 1965, 1966) concerned with Lorentz and with relativity theory are highly recommended. I have also had the privilege of seeing some unpublished papers by S. Goldberg and by R. McCormmach, which it is hoped, will soon be more widely available.† Some of the philosophical aspects connected with the aether and relativity are developed in the important works of M. B. Hesse (1965) and A. Grünbaum (1963).

I should like to express my thanks to the directors and librarians of the institutions which assisted me in my research, among them the History and Philosophy of Physics Center at the American Institute of Physics, New York; the Institution of Electrical Engineers, London; the Algemeen Rijksarchief in the Hague; and the Deutsches Museum in Munich. I am also indebted to Messrs. Taylor and Francis for permission to reprint the Michelson–Morley paper from the *Philosophical Magazine*, to Cambridge University Press to reprint the Larmor selections, and to Michael G. Heaviside of the Oliver Heaviside Educational Foundation for permission to reprint the Heaviside selection. Finally I should like to express special appreciation to Melba Phillips and to Dudley Shapere for their critical reading of the commentary, to William Lycan for translation help, to Carl Dolnick for mathematical assistance, to Jan Jones for secretarial help, and to my wife for assistance with the manuscript. I am also indebted to the National Science Foundation for support of the research for this book.

K. F. S.

---

† R. McCormmach, "H. A. Lorentz and the Electromagnetic View of Nature" (mimeo); "Einstein, Lorentz, and the Electron Theory" Historical Studies in the Physical Sciences, II. (in press); and S. Goldberg "In Defense of Ether: The British Response to Einstein's Special Theory of Relativity, 1905–1911" (mimeo); "The Lorentz Theory of Electrons and Einstein's Theory of Relativity" *Am. J. Phys.*, Oct. 1970.

PART 1

CHAPTER I

INTRODUCTION: THE FUNCTIONS OF THE AETHER

THROUGH the nineteenth century, various ideas of the aether dominated much of optical and electromagnetic theory. Though aether theories had been proposed in previous centuries, it was primarily through the development and acceptance of a powerful wave theory of light that more and more attention became focused on the nature of the optical medium. At first aether theories were attempts to explain mechanically various optical laws and optical phenomena. Later, with the development in the latter part of the nineteenth century of Maxwell's electromagnetic theory of light, a number of scientists tried to formulate mechanical aether theories adequate to explain Maxwell's theory, and derivatively, physical optics.

The aether approach offered a means of applying the elegant Lagrangian and, later, the Hamiltonian forms of mechanics to optics and electromagnetism. It seemed that a "unified field theory", to use more current jargon, might be the result of careful research into the nature of the aether. Maxwell's theory, at least in the British Isles, inspired some scientists to point their research in this direction. Among them I might mention Oliver Heaviside, George Francis Fitzgerald, and Sir Joseph Larmor. Larmor, the author of the influential *Aether and Matter* (1900), surveyed the field of aether research in 1907 and wrote in a statement that is typical even though it came in the twilight of the aether approach:

> It [the aether] must be a medium which can be effective for transmitting all the types of physical action known to us; it would be worse than no

solution to have one medium to transmit gravitation, another to transmit electric effects, another to transmit light and so on. Thus the attempt to find out a constitution for the aether will involve a synthesis of intimate correlation of the various types of physical agencies, which appear so different to us mainly because we perceive them through different senses.

It should also be noted, with reservations to be explored in Chapter V of this Commentary, that such a "unified field theory" would have been a *mechanical* theory, thus accomplishing a complete unification of physics that has often been an inspiration and goal to physicists from Oersted and Faraday to Einstein and Wheeler.

The aether also played a most significant role in the evolution and revolution of ideas of space and time. As the medium of optical, and then electromagnetic activity, the aether was assumed by many to constitute the absolute frame of reference in which the equations of the optical aether and Maxwell's equations would have their simplest form. As the Earth was clearly in motion about the Sun, effects of this motion were conceivable, depending on the aether theory held, which would be experimentally accessible. But serious difficulties appeared in connection with a consistent aether theoretic explanation of stellar aberration, the partial dragging of light waves by moving transparent media, and the null result of the Michelson interferometer experiment. Eventually, to accommodate the last result, Fitzgerald and Lorentz proposed a contraction hypothesis in which the length of an object depended on its velocity through the aether. Soon after this, Lorentz and Larmor developed more radical hypotheses by which certain compensating effects, including an alteration of time, cooperated in eliminating most aether wind effects. Finally, Einstein, knowing of the null effects of various aether drift experiments and of the dependence of electromagnetic induction of *relative* velocities only, and cognizant of some of Lorentz's ideas, took a most revolutionary step. He articulated a "principle of relativity" and built a theory on it which showed that the Galilean and Newtonian ideas of space and time were in error, and that a simple and consistent electromagnetic theory would require the elimination of the notion of an aether rest frame and require a new understanding of simultaneity and the way in which the spatial and temporal aspects of processes were connected. Though this book does not present anything like a careful analysis of Einstein's special theory of relativity, it does develop the aether concept to the point where it can be shown, with the help of *some* of Einstein's theory, why the aether was eliminated from physics.

It is hoped that this book, in addition to informing scientists about some of the discarded foundations of electromagnetism and special relativity, might have some effect in stimulating a renewed interest in the history of the aether by historians of science. It is also hoped that it might provide a number of philosophers of science with an insight into the sophisticated complexities of mechanical explanations and mechanical "models", the relation between theory and experiment, theory change, *ad hoc* modifications of theories, and scientific revolutions. The last topic has been the focus of considerable interest following the publication of T. S. Kuhn's *The Structure of Scientific Revolutions* (1962) and some essays by P. K. Feyerabend (e.g. 1962, 1965a, 1965b). One of the *implicit* theses developed in the present book is a critique of those views, such as Kuhn's, which argue that logic and experimental evidence are of little weight in the process of theory replacement. It has become more and more evident to me in doing the research for this book that Kuhn's claims—and also to some extent Feyerabend's—regarding the lack of rationality and experimental control in the development of science constitute serious oversimplifications of the history of science. Point by point refutations of Kuhn and Feyerabend are not the function of this essay, but it is hoped that a careful and historically accurate account of the rise and fall of various nineteenth-century aether theories will constitute the ground on which one can be built.[†]

[†] I do not want either to imply that Kuhn's and Feyerabend's views of the nature of theory replacement are identical, nor that I agree with none of

their claims. For an example of their differences consider that Kuhn's view of scientific development argues that *one* paradigm at a time characterizes a science after the pre-paradigm, pre-scientific state, and that this paradigm is instrumental in a dialectical process of creating its sole nemesis and consequently its own downfall. On the other hand, Feyerabend's view of proper scientific development has it that there are sets of partially overlapping but mutually inconsistent theories, each element of which competes for the allegiance of a scientist, their relative merits apparently depending on their relative ability to survive falsification. The last point is, however, somewhat questionable within Feyerabend's philosophy of science, because of the tight connections and influences between theory and observation, but he does claim in various places (1965b) that his approach is built on Popper's work, and unless it totally transcends its Popperean foundations, falsifiability and falsification must play a central role in theory evaluation. See Sir Karl Popper's important (1959) monograph. Since this manuscript was completed, I have formulated an account of the logic of scientific development, and have applied it to the case of theory competition between Lorentz' absolute theory of the electrodynamics of moving bodies (discussed below in Chapter VI) and Einstein's relativistic theory. See my (1970) essay, "Outlines of a Logic of Comparative Theory Evaluation with Special Attention to Pre- and Post- Relativistic Electrodynamics," in *Minn. Stud. in Phil. of Sci.*, 5, ed. R. Stuewer, University of Minnesota Press, Minneapolis.

CHAPTER II

THE HISTORICAL BACKGROUND OF THE NINETEENTH-CENTURY AETHER THROUGH YOUNG AND FRESNEL

THE aether played an important role in natural philosophy and physics from ancient times to the beginning of the twentieth century. Conceived in its most general terms as a "thin subtile matter, or medium, much finer and rarer than air", it was believed by most to fill the celestial regions, and by some to also pervade the air and even solid bodies. Some natural philosophers considered it a fifth element, or "quintessence", not reducible to combinations of the four elementary substances of earth, air, water, and fire. The question as to whether the aether was "ponderable", or subject to the forces of gravity, was also occasionally debated, the most recent proponent of its ponderability being Max Planck, the founder of quantum mechanics.[†]

In its long history, the aether has had a number of different tasks assigned to it, such ascriptions occasionally leading scientists to perhaps multiply aethers beyond necessity.[‡] For Descartes, the aether was a form of matter that was transparent and which filled those regions where the matter of the Earth and Sun were not. Such an aether was required by Descartes since "extension" was the essential property of matter or the *Res Extensa*, and the Universe, in order to exist physically, was required to be a *plenum*.

[†] Planck's theory of the ponderable compressible aether is briefly discussed in Chapter III of this book.

[‡] See Whittaker (1960), I, pp. 99–100, for a discussion of the problem of one *versus* many aethers.

Not much later Isaac Newton utilized the aether for various other purposes. For Newton an aether was required to explain the transmission of heat through a vacuum, and Newton also attempted, unsuccessfully, to develop an explanation of his inverse square law of gravitation using the aether. In addition, Newton employed an aether in his optical theory, not as a medium through which light could be propagated like sound is propagated through air, but rather as a medium which could interact with light corpuscles to produce refraction phenomena and Newton's rings.

In direct opposition to Newton's approach, the central task of the nineteenth-century aether, with which this book is primarily concerned, is to provide a medium for the propagation of light waves. But wave theories of light and theories of the optical aether had been developed before and during Newton's period by Robert Hooke in his *Micrographia* (1665) and by Christiaan Huygens in his well-known *Traite de la Lumiere*... (1690). It will be important for reasons to be made clearer later to discuss one of these theories briefly, before we consider the work of Young and Fresnel. Since Hooke's work was rather vague and unquantitative, and because Hooke, unlike Huygens, had little influence on nineteenth-century aether theories, I shall begin with Huygens' contributions and with Newton's criticisms of them.

1. Huygens

Huygens worked within the Cartesian tradition of physics, though he did differ with Descartes on certain points, notably on the finite velocity of light propagation. Huygens proposed that light must be a mechanical motion conveyed from a luminous body to the eye. For Huygens (1690), "in the true philosophy... one believes all natural phenomena to be mechanical effects.... We must admit this or else give up all hope of ever understanding anything in physics." As we shall see later, a belief in the fundamental character of mechanics is a common assumption held by almost all aether theorists.

Since light is motion and since two light beams could cross one another at any angle without disturbing one another, Huygens did not believe light could be the motion of material particles translated from the luminous object to the eye. Rather, Huygens (1690) argued: "Light is propagated in some other manner, an understanding of which we may obtain from our knowledge of the manner in which sound travels through air".

Huygens, accordingly, proposed a medium he called the aether, which he conceived of as a dense collection of very small, very rigid elastic spheres, through which light was propagated. These spheres filled all space and even penetrated into "solid" material bodies through their hidden porous structure. Furthermore, Huygens proposed that the aether and the matter interacted so as to affect the velocity of the light in the bodies, since in an intermingled state the total elasticity of the medium could be considered diminished thus retarding the velocity of propagation of a wave through it. Explanations of refraction and even double refraction were founded on these ieas, and worked out by Huygens using his own recently discovered principle of secondary wave propagation.

For Huygens, light waves were very much like sound waves, for even though they were propagated with a considerably higher velocity than sound, they were strictly longitudinal in form. Isaac Newton, whose optical investigations both preceded and postdated Huygens' work, could not accept a wave theory of light as he was unable to see how well-defined rays and sharp shadows could be explained by such a theory. Newton favored a corpuscular theory of light, such as Huygens had rejected, which did not require an aether for quite the same reasons that Huygens' theory did, though Newton, as noted above, did use an aether for other purposes in his optics. Later, in 1717, Newton felt confirmed in the wisdom of his rejection of the wave theory when he discovered that a ray of light which had been obtained by double refraction differed from ordinary light in that the former possessed a directional orientational property the latter did not have. Newton talks of the ray obtained from double refraction as having "Sides", such as may be possessed by a rectangle but not by a circle. (Later Malus termed this property "polarization".) Newton was quite convinced that a wave theory of light could not explain such a property, though a corpuscular theory might, as corpuscles themselves could have sides.

Such objections, together with Newton's growing authority in physics, brought hard days on the proponents of the wave theory of light during the eighteenth century, and accordingly on the development and acceptance of aether theories in which the aether functioned as the light medium. The rejection of the aether during the eighteenth century was apparently also aided by the influence of the "philosophical" or methodological preface which Roger Cotes introduced into Newton's second edition of the *Principia* in 1713. In this preface Cotes polemized against Descartes and the Cartesians who:

> When they take a liberty of imagining at pleasure unknown figures and magnitudes, and uncertain situations and motions of the parts, and moreover, of supposing occult fluids, freely pervading the pores of bodies, endued with an all-performing subtility, and agitated with occult motions, . . . run out into dreams and chimeras, and neglect the true constitution of things, which certainly is not to be derived from fallacious conjectures, when we can scarce reach it by the most certain observations.

Cotes distinguished natural philosophers into three camps: (1) the Newtonians, who founded their science on experiments and observations, and who subscribed to action at a distance and a "void" in their gravitational theory, (2) the Aristotelians and Scholastics whom Cotes summarily dismissed, and (3) the Cartesians, who fill the void with vortices and subtle matter to the detriment of true scientific philosophy. It is easy to see what might have happened to an aether theory approach as such a positivistic philosophy of science became widespread during the eighteenth century.[†]

2. Young

The authority which Newton's theories eventually came to exercise over physics by the beginning of the nineteenth century is well displayed in comments in the writings of Thomas Young, a physician who began to work on sound and light in the closing years of the eighteenth century. In 1800 Young published his first thoughts on a wave theory of light as a small part of a paper discussing some experiments on sound. In the section of this paper titled "Of the Analogy between Light and Sound", Young proposed a wave theory of light rather similar to Huygens' theory, except that the difference in velocities of light in media was ascribed to differences in the aether's *density* rather than to differences in rigidity.

About two years later Young (1802) worked out his thoughts in somewhat more detail. Apparently in an attempt to win his ideas a fair hearing from the Newtonians, Young quoted Newtonian "scripture" in support of the hypotheses of his wave theory. The spirit of the times is perhaps aptly caught in Young's (1802) statement that:

> Those who are attached, as they may be with the greatest justice, to every doctrine which is stamped with the Newtonian approbation, will probably be disposed to bestow on these considerations so much the more of their attention as they appear to coincide more nearly with Newton's own opinions.

Young collected passages from Newton's scattered writings on optics and the aether in support of the four hypotheses of his own aether and wave theory. These hypotheses, which indicate quite

---

[†] The influence of Cotes' preface is discussed by Whittaker (1960), I, pp. 30–31.

clearly the connection between a theory of optics and an aether at the beginning of the nineteenth century were:

1. A luminiferous aether pervades the universe, rare and elastic in a high degree.
2. Undulations are excited in this aether whenever a body becomes luminous.
3. The sensation of different colors depends on the different frequencies of vibrations excited by light in the retina.
4. All material bodies have an attraction for the aethereal medium, by means of which it is accumulated in their substance and for a small distance around them, in a state of greater density but not of greater elasticity.

Young deduced several propositions from these hypotheses concerning the common velocity of light waves in a medium, and spherical form of the wave, the refractive capacity of a medium as a function of its aether density, and a principle of interference of waves. Young continued his work in the next few years and in a volume published in 1807 (Young 1807) he discussed his famous two-slit interference experiment which is so often cited as proving the existence of waves of light. Young also attempted provisional explanations of inflection or diffraction, and did some excellent work on the colors of thin plates and Newton's rings. Later, in 1809, he defended Huygens' theory of double refraction, along with several modifications of his own, against Laplace's corpuscular theory of double refraction.

Though Young's experiments were brilliantly conceived and executed, the arguments which he gave in support of his *theory* did not go much beyond what Huygens had accomplished. Young apparently was somewhat deficient in training in mechanics and higher mathematics, and was not able to bring the sophisticated theoretical developments of recent science to work in his favor.[†]

[†] See Crew's (1900) brief biography of Young in which he suggests this.

The Newtonian school was still exceedingly strong in the beginning of the nineteenth century, and to say that Young's theory did not attract many followers would perhaps, on the evidence of his biographer, be understating the situation.† Even after Young's fundamental papers on the wave theory and its experimental foundations had appeared, Herschel and Laplace continued to develop optics in the corpuscular manner, and to ignore Young's contributions.

There were some reasons for this other than simple scientific inertia. In the years 1808–10 E.-L. Malus performed a number of experiments on the intensity of light that was reflected from a transparent body's surface. Malus analyzed the light using a double refracting crystal of Iceland spar and noted that light reflected from the surface of transparent media possessed the same property of having "Sides" which Newton had noticed in connection with doubly refracted rays. Malus gave the name "polarization" to this property, and attributed it, on the basis of a particle theory of light, to light corpuscles having their sides all turned in the same direction, much as a magnet turns a series of needles all to the same side. Subsequent to Malus' publication, the French physicist Biot developed a more complex corpuscular theory of polarization.

In England in the years 1814–19, David Brewster conducted several experiments, some of them similar to Malus', and obtained results which were of considerable significance to the future of optics. Among these results was a formula connecting the angle of complete polarization of the reflected ray with the refractive index of the media—a relation which Malus had sought but could not determine. Brewster also discovered the existence of biaxal crystals in which there were two axes along which double refraction did not occur, rather than one axis as in Iceland spar. The immediate effect of Brewster's discovery was to call into serious question Huygens' analysis of double refraction and the wave theory of

† Whittaker refers to Peacock's *Life of Young* in recounting the incident in which Young's pamphlet replying to a scathing attack on his wave theory in the *Edinburgh Review* only sold one copy.

light in general since Huygens' construction no longer sufficed to account for the refraction in the more complex biaxal crystals. Brewster also empirically obtained what have come to be called Fresnel's sine and tangent laws, about which I shall have more to say below.

3. Fresnel

It was clear that polarization was a problem which was exceedingly difficult to explain on the basis of the wave theory, as long as light was conceived on the analogy with sound. Sound waves, as *longitudinal* waves, could not account for the "sidedness" displayed in polarization phenomena. Young, purportedly reflecting on Brewster's experiments and on the results of an experiment carried out in France by Arago and Fresnel,[†] was the first to suggest a possible explanation of polarization on the basis of the wave theory. In a letter to Arago dated 17 January, 1917, Young proposed that if light waves were conceived of as *transverse* waves, they could admit of polarization. Not long after, in another letter to Arago, Young compared light waves to the motions of a cord which has one of its ends agitated in a plane.

Arago showed this letter to Fresnel who at once seized upon the hypothesis of transverse waves as one with which he could explain polarization. Subsequently, Fresnel made this hypothesis the basis of his most influential dynamical theories of double refraction and reflection and refraction.

In the years between 1814 and 1818, Fresnel had already made very important contributions to the wave theory of light. His great memoir on diffraction (Fresnel, 1826) was developed gradually during these years and is worked out on the basis of the older

† This is the experiment in which two pencils of light polarized in planes at right angles to one another cannot be made to interfere under any condition of path-length difference. The results were not published until 1819 though the experiment had been done several years earlier.

longitudinal wave theory of light. But since this inquiry was essentially "kinematical" and not concerned with the true motions of the aethereal medium, the shift to an aether which would support transverse vibrations did not vitiate his diffraction theory.† Fresnel also developed in the year 1818 another theory which was not strictly dependent on the structure of the medium and the type of wave it would support. This is Fresnel's famous explanation of aberration phenomena and includes the derivation of his partial dragging coefficient. I shall have more to say about this inquiry in the next chapter.

I now turn to consider, somewhat sketchily, Fresnel's two important dynamical theories of light. My intention here is not to present an adequate account of Fresnel's dynamical theories, but rather to emphasize the important but unsatisfactory aspects of these so as to prepare the reader for the more adequate aether theories of the later chapters. Fresnel never actually worked out an acceptable *mechanical* theory of light, though this seems to have been his intention, and his accounts are at best quasi-mechanical or quasi-dynamical attempts at the analysis of wave motions of the aether. Nevertheless these attempts were of the greatest significance because of their plausibility and simplicity, their agreement with experiment, and their ability to predict new experimentally confirmable phenomena. The fact that Fresnel could so effectively systematize optics from the point of view of the transverse wave theory soon resulted in the almost complete acceptance of the wave theory and the consequent rejection of the corpuscular approach.

Fresnel's first attempt at a dynamical theory of the aether medium focused on the problem of double refraction which had been raised anew by Brewster's biaxal crystals. The result of Fresnel's (1821) inquiry was a theory of double refraction which was characterized in 1834 in a report to the British Association for the Advancement of Science in the following eulogistic terms:

† Fresnel's diffraction theory was not given an appropriate dynamical basis until G. G. Stokes' essay (1849) "On the Dynamical Theory of Diffraction".

> The theory [of double refraction] of Fresnel to which I now proceed,—and which not only embraced all the known phenomena, but has even outstripped observation, and predicted consequences which were afterwards fully verified—will I am persuaded, be regarded as the finest generalization in physical science which has been made since the discovery of universal gravitation.[†]

Fresnel's theory is an "elastic solid theory" in that it assumes that the aethereal medium is so constituted as to permit transverse waves to be propagated through it. It is not a continuum but, like Huygens' aether, consists of a huge number of very small aether molecules with forces acting between them. There is nothing inconsistent with using this type of molecular hypothesis as the basis of a general mechanical theory of the aether, and it was later employed by Cauchy and by Green, whom I shall consider in Chapter IV. Fresnel did not pursue a true mechanical approach, however, but rather introduced hypotheses additional both to the mechanical laws of motion and to the force functions expected in an elastic medium.[‡]

One of Fresnel's additional hypotheses was innocuous and concerned the relation between the vibrations of the medium and the direction of the plane of polarization; it assumed that the vibrations of polarized light were at right angles to the plane of polarization. Fresnel's three other assumptions, however, were somewhat artificial and even inconsistent with a true elastic solid theory. For example, his second hypothesis, that the elastic forces produced by the propagation of a transverse plane wave were equal to the product of the elastic force produced by the displacement of a single molecule of the aether multiplied by some constant which is independent of the direction of the wave, is not true in the case of a mechanical elastic solid. In elastic solids, the elastic forces

---

[†] The report was authored by Humphrey Lloyd (1834). The various phenomena that were accounted for and predicted *de novo* are given in Lloyd's essay.

[‡] The assumptions of Fresnel's theory were analyzed by E. Verdet (1869), on whose work the following account of Fresnel's double refraction theory is based.

are functions of the particle's displacement *relative* to its neighbors, and the implied restriction of the elastic force to act only along the line of displacement is also false. In still another hypothesis, Fresnel stipulated that only the component of the elastic force parallel to the wave front was to be considered effective in the propagation of a light wave. This amounts to an *ad hoc* elimination of the longitudinal wave that should also be propagated in a disturbed elastic solid. (The problem of the longitudinal wave will be considered in detail in Chapter IV.) Finally, Fresnel's fourth hypothesis asserted that the velocity of a plane wave is proportional to the square root of the effective component of the elastic force developed by the wave. But this hypothesis had for its foundation only the analogy that such a relation holds in the case of transverse vibrations of a stretched string.[†]

Fresnel's theory of double refraction is clearly not a reduction of optics to mechanics, and Fresnel himself expressed concern about the security of foundation of his hypotheses. Nevertheless, he thought them warranted by the consequences which could be drawn from them, and draw them he did. He obtained acceptable explanations of double refraction, solving the problem of biaxal crystals and showing that Huygens' construction for uniaxal crystals was a special case of his own more general wave surface. It was subsequently determined by Hamilton that the Fresnel theory would imply the unanticipated phenomena of conical refraction, which was then sought for and found. At the time all available experimental tests of Fresnel's theory showed striking agreement between theory and experiment. Later, however, more precise optical experiments showed Fresnel's wave surface to be only a very close approximation to the actual surface.[‡]

[†] These criticisms are in part based on Verdet's (1869) work and Preston's (1895) account, pp. 318–23, but are paralleled by similar ones in Whittaker (1960), I, p. 119.

[‡] See Whittaker (1960), I, pp. 121–2, for more detail on this topic.

Two years after he had developed the theory of double refraction, Fresnel (1832) proposed another quasi-dynamical theory: this one to take account of reflection, refraction, and the polarizing properties of the surfaces of transparent media. In this theory Fresnel proceeded similarly to the way he had two years earlier. He made use of whatever dynamical principles he could use, e.g. the conservation of *vis viva;* he violated other mechanical principles, such as the continuity of the normal component of the aether displacement across an interface—this resulted in a covert inconsistency with his other boundary conditions. Fresnel also added the unwarranted but plausible assumption of Young that refraction depended on the differences in density and not in rigidity. He obtained valuable results though, among them derivations of the important laws connecting the relative amplitudes of the reflected and incident waves. If the wave was polarized in the plane of reflection he obtained:

$$\frac{A_i}{A_r} = \frac{\sin(i-r)}{\sin(i+r)}$$

in which i is the angle of incidence of the normal of the plane wave and r the angle of refraction of the normal of the refracted waven. If the wave were considered to be polarized perpendicular to the plane of reflection, then the amplitude ratios became:

$$\frac{A_i}{A_r} = \frac{\tan(i-r)}{\tan(i+r)}.$$

In spite of the dynamical insecurity of the foundations of these theories of double refraction and reflection, it would be unfair to criticize Fresnel for failing to avail himself of the proper mechanical bases. For at the time when Fresnel was constructing these theories, the dynamics of an elastic solid were only beginning to be developed and were not worked out in any satisfactory form. In fact, it was due in part to Fresnel's groping attempts at such theories that the French mathematical physicists Navier, Poisson, and Cauchy were stimulated to perfect an adequate mechanical theory of the motions of an elastic solid. This is a topic to which I shall return in Chapter IV.

This historical introduction, then, brings us somewhat into the nineteenth century, and relates how the wave theory of light together with its elastic solid aether developed in the early quarter of the century of our concern. But with the exception of a brief aside, I have ignored one basic problem which confronted the wave theory of light from 1727 on, and which is now considered to have been adequately solved only by Einstein's special theory of relativity in 1905. This is the problem of aberration, first discovered in starlight by Bradley, and later experimentally examined in both stellar and terrestrial cases by Arago, Airy, and Michelson and Morley. In the next Chapter I shall consider this problem and its connection with the aether of the wave theory of light. I shall return to the dynamical problems of the optical aether in the following chapter where I shall consider in some detail the elastic solid theories of Green, MacCullagh, and Lord Kelvin.

CHAPTER III

ABERRATION FROM BRADLEY TO MICHELSON AND MORLEY

IN THE previous chapter I outlined in fairly broad terms the development of the elastic solid aether in the early nineteenth century. I now move to more detailed investigations of the nineteenth-century aether theories. Following the plan of most nineteenth-century aether monographs, I shall begin a close scrutiny of the aether by examining the connections that were thought to exist between aether and ponderable matter. Later I shall discuss the hypothesized nature of the aether in free space. I shall not consider all of the aether-matter connections in this chapter, but will primarily be concerned with the aberration observations and experiments performed during the eighteenth and nineteenth centuries.†
Problems of dispersion, in which the aether interacts with matter, will be referred to near the conclusion of the next chapter. There are, of course, certain problems which arose in connection with light and matter, such as the photoelectric effect, which never obtained an explanation on the basis of an aether theory.

Aberration problems are concerned with the effect of the motion of ponderable matter on aether phenomena, such as the velocity of light waves in the aether as viewed from moving ponderable

† The alteration in the observed frequency of the light waves produced by relative motion, and at one time it was thought absolute motion, will not be discussed here, as it does not seem to have been very important in the development and replacement of aether theories. This phenomenon, known as the Doppler effect, is different in the aether and relativity theories, however, and would have to be discussed in any thorough comparison of these two approaches.

matter, or the effect of the motion of ponderable matter on light moving within it. Throughout most of the nineteenth century, all known aberration phenomena were optical. Toward the close of the century, however, with the development of Maxwell's and Lorentz' theories, some electrical and magnetic aberration experiments were performed. These were occasioned by Hertz' and Lorentz' theories and were satisfactorily accounted for by Lorentz' theory.

1. Bradley

In the years 1725–6 Samuel Molyneux, with some assistance by James Bradley, the Professor of Astronomy at Oxford, attempted to carry out a careful experiment designed to detect traces of the annual parallax of the "fixt Stars". Beginning on 3 December, 1725 and intermittently over a period of twelve months, measurements were made of the position of the star γ *Draconis*, and an apparent causal influence of the Earth's motion on the direction of the star was observed. It was not, however, the sort of result which the two astronomers had been expecting. Bradley (1728) noted that: "this sensible alteration the more surprised us, in that it was the contrary way from what it would have been, had it proceeded from the annual parallax of the Star". The apparent displacements of the star, instead of being directed towards the Sun as expected, were in a direction perpendicular to the earth's orbit.

In 1727–8, with an instrument which was less constrained in movement than Molyneux's, Bradley alone repeated the observations, this time being able to observe several stars. He again found the same perplexing phenomenon. Larmor (1900) reports that Bradley was led to the true explanation of this phenomenon by the "casual observation of a flag floating at the masthead of a ship; when the ship changed its course, the flag flew in a different direction". Bradley does not recount this particular story in his paper, however, which he published in the *Philosophical Transactions* (1728), but rather states simply:

> At last I conjectured, that all the *Phaenomena* hitherto mentioned, proceeded from the progressive Motion of Light and the Earth's annual Motion in its Orbit. For I perceived that if Light was propagated in Time, the apparent Place of a fixt Object would not be the same when the Eye is at Rest, as when it is moving in any other Direction, than that of the Line passing through the Eye and Object; and that, when the Eye is moving in different Directions, the apparent Place of the Object would be different.

Bradley formulated a mathematical expression relating the apparent displacements to the velocity of the earth and the velocity of light:

> And in all Cases, the Sine of the Difference between the real and visible Place of the Object, will be to the Sine of the visible Inclination of the Object to the Line in which the Eye is moving, as the Velocity of the Eye to the Velocity of Light.

How this follows can be seen from Fig. III.1. The star is at A with the earth moving from B towards C. A telescope, were the Earth stationary, would be sighted parallel to line CA, but, since the Earth is in motion, the telescope must be sighted along a line parallel to BA. From the law of sines it follows that $\sin \alpha / CB = \sin \beta / AC$. Since $DB : AC :: v : c$, where v is the orbital velocity of the Earth and c the velocity of light, we have $\sin \alpha : \sin \beta :: v : c$, which is Bradley's relation. If the angle α is small, and it is since v is much smaller than c, then one can write to a close approximation $\tan \alpha = v/c$. This is usually the way in which the law of aberration is expressed in contemporary texts. The ratio v/c is often referred to as the *aberration constant*.

Bradley explained his findings in terms of the corpuscular theory of light, on whose basis the addition of velocities is physically very plausible. One only need assume that the corpuscles are not affected by the Earth's gravitational attraction. On the basis of the wave theory, the true path of the light is more difficult to explain, for the explanation apparently has to involve the assumption that the

Earth's motion through the aether medium does not affect the motion of the medium. Thomas Young (1804) actually made this suggestion in connection with an explanation of aberration when

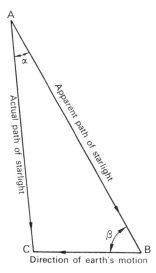

FIG. III.1. Stellar aberration.

he wrote: "Upon considering the phenomenon of the aberration of the stars I am disposed to believe that the luminiferous aether pervades the substance of all material bodies with little or no resistance, as freely perhaps as the wind passes through a grove of trees."

2. Fresnel

In the beginning of the nineteenth century the French physicist François Arago reasoned, on the basis of the corpuscular theory of light, that the aberration of light in an optically dense medium, such as in a glass prism, would be different if the incident starlight were passed through the prism in the same direction as the Earth's motion than if it were passed in an opposite direction. Arago's experiments to test this hypothesis were performed in 1808–9 but gave a null result. Though their accuracy has since been questioned, it appeared at the time that a most peculiar phenomenon was occurring. Though it was clear that the motion of the Earth affected the direction of apparent propagation of incoming light from a star, when the same light was sent through a refracting medium, the medium exerted no *additional* aberrational effect on it. This seemed to imply that aberration was and yet was not operative. Some years later, after Fresnel had made his initial contributions to the wave theory of light, Arago wrote to him telling him of these experiments and of his inability to conceive of a reasonable explanation on the basis of corpuscular theory. He asked Fresnel whether an explanation in terms of the wave theory might be possible.

Fresnel replied to Arago's query in a letter which was subsequently published in the *Annales de Chemie*... (1818). In this letter Fresnel was able to formulate a simple and elegant explanation of Arago's results on the basis of the wave theory of light; an explanation which not only accounted for aberration effects then known, but which was subsequently confirmed in a number of different ways throughout the nineteenth century.

Fresnel began his letter to Arago by considering possible alternative explanations. The corpuscular interpretation seemed improbable to Fresnel for the reason that it would require, as Arago had suggested, that a radiating body would have to impart infinitely many different velocities to light corpuscles, and that the corpuscles would affect the eye with only one of those velocities. Complete aether drag was also ruled out, for though it would explain the null effects of the Earth's motion on *refraction* phenomena, it apparently could not explain Bradley's aberration phenomena. Fresnel accepted Young's idea that aberration phenomena, to be explained on the basis of the wave theory, would have to allow the aether to

ABERRATION FROM BRADLEY

pass freely through the earth, or at least not disturb the aether's motion in the atmosphere where aberration observations had been carried out.

Fresnel proposed to account for both Arago's result and aberration phenomena by supposing a *partial* aether drag in which transparent bodies with refractive indices greater than a vacuum (in which the index $n = 1$) were conceived to have a greater aether density within them, and that only the aether density which constituted an *excess* over and above the aether density in the vacuum would be completely carried along by the moving body.

Fresnel's argument to support this point and to derive his partial dragging coefficient is somewhat analogical and not very convincing. He supposes, like Young before him, and as he himself does in his later paper on reflection and refraction, that the index of refraction (n), the velocity of light (c), and the densities of the aether ($\triangle$) in empty space and within the body are related by:

$$\frac{c}{c_b} = \frac{n_b}{n} = \frac{n_b}{1} = \frac{\sqrt{\triangle_b}}{\sqrt{\triangle}} \qquad (3.1)$$

the b subscript distinguishing the c, n, and $\triangle$ within the transparent body. In a moving body in which $n > 1$, only the excess of the aether is considered dragged. Fresnel gives the following argument in support of his reasoning that if this is so, then only a partial augmenting of the *velocity* of light *in* the moving medium will occur:

> By analogy it would seem that when only a part of the medium is displaced, the velocity of propagation of waves can only be increased by the velocity of the centre of gravity of the system.
>
> The principle is evident in the case where the moving part represents exactly half of the medium; for, relating the movement of the system to its centre of gravity, which is considered for a moment as fixed, its two halves are travelling away from one another at an equal velocity in opposite directions; it follows that the waves must be slowed down in one direction as much as they are accelerated in the other, and that in relation to the centre of gravity they thus travel only at their normal velocity of propagation; or, which amounts to the same thing, they share its movement. If

> the moving portion were one quarter, one eighth or one sixteenth, etc., of the medium, it could be just as easily shown that the velocity to be added to the velocity of wave propagation is one quarter, one eighth, one sixteenth, and so on, of that of the part in motion—that is to say, the exact velocity of the centre of gravity; and it is clear that a theorem which holds good in all these individual instances must be generally valid.
>
> This being established, and the prismatic medium being in equilibrium of forces (tension) with the surrounding ether (I am supposing for the sake of simplicity that the experiment is conducted in a vacuum), any delay the light undergoes when passing through the prism when it is stationary may be considered as a result solely of its greater density.... .

By Fresnel's supposition, only the *excess* aether density above the vacuum's density is dragged by a moving transparent body, i.e. $(\triangle_b - \triangle)$ which by (3.1) above equals $\triangle_b(1 - 1/n_b^2)$. In accordance with the reasoning developed in the quote above, the *increase* in the velocity of the light within the moving media will be $v(1 - 1/n_b^2)$. The factor $(1 - 1/n^2)$ is Fresnel's famous "partial dragging coefficient", variously called Fresnel's convection coefficient or the coefficient of entrainment. Regardless of what one may think of the argument by which it was deduced, the coefficient is of the greatest importance in aberration theory. It was noted by Lord Rayleigh as late as 1892 in connection with various aberration problems that: "It is not a little remarkable that this formula [i.e. the convection coefficient] and no other is consistent with the facts of terrestrial refraction, if we once admit that the aether in the atmosphere is at absolute rest."

In his 1818 letter Fresnel showed that his partial dragging hypothesis would adequately explain Arago's result, and, moreover, that it would also predict that filling an aberration detecting telescope with water would have no effect on the observed aberration. Such an experiment had been proposed in the previous century by Boscovich, but it was not carried out until 1871 by Airy, who did obtain Fresnel's predicted result.

Rather than presenting a reconstruction of the way in which Fresnel shows that his formula accounts for Arago's experiment—and his reasoning is not very explicit on this point, as can be seen from the Fresnel selection, pp. 128–31, I propose to give instead an example which is closer to Fresnel's water-filled telescope or microscope case, but which can easily be seen to extend to the analysis of a moving prism. This example is a modification of one of H. A. Lorentz' (1901).

We consider light from a star impinging on a moving system as is shown in Fig. III.2. The tube *ABCD* is empty (a vacuum) and

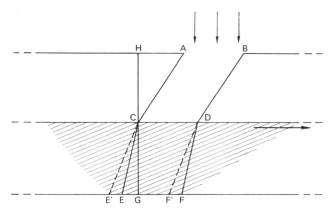

FIG. III.2. Stellar aberration in a vacuum and in a moving transparent medium. (After Lorentz, 1901.) The shaded area is filled with glass.

the tube *CDEF* is filled with glass with an index of refraction $n > 1$. The light from the star follows the apparent path parallel to *AC* and *BD* and strikes the glass at *CD* at angle of incidence i. Let us suppose now, for the moment, that there is *no* influence of the Earth's motion on *refraction* phenomena. If this were so, then simply taking the angle i as given and applying Snell's law we get:

$$n \sin r = \sin i, \tag{3.2}$$

the index of refraction of the vacuum being equal to 1. Let v be the velocity of the Earth through the aether, c the velocity of light in the vacuum, and c_b the velocity of light in the glass. By Bradley's

aberration experiment just discussed, we can set $\tan i = v/c$, or for small angles:
$$\tan i = \sin i = v/c. \tag{3.3}$$

From (3.2) and (3.3) we then have $\tan r = v/nc$, since r is even smaller than i, and accordingly segment length $EG = lv/nc$, if $CG = l$.

Whatever the length of EG is, it represents a measure of the effect of aberration within this system. We have now calculated its length assuming that the motion of the Earth exercises no effect on refraction phenomena, though we have included the effect of the Earth's motion on aberration in a vacuum.

But this calculation above is inconsistent with the assumption of an absolutely stationary aether through which the glass is moving, unless there are effects which compensate for the aether wind which is, by our hypothesis that is supported by the aberration effect in the vacuum, blowing through the glass with a velocity equal to $-v$. Let us see what the effect of the aether wind within the glass would be if the aether passed through the glass as freely "as the wind passes through a grove of trees". We now consider the true direction of the light (as we did in calculating the angle of aberration for the vacuum) as perpendicular to the interface CD. Since the light moves, on this assumption, with velocity c_b in the vertical direction and with velocity $-v$ in the horizontal direction we obtain the two new points, $E'G'$, where it intersects with the base line of Fig. III.2. From similar triangles it follows that:
$$E'G : CG = v : c_b = v : c/n \tag{3.4}$$

whence $E'G = lnv/c$. The difference between EG and $E'G$ represents the difference between the outcome of Arago's experiment and the predicted outcome of a wave theory *without* Fresnel's partial dragging coefficient. If Fresnel's hypothesis is correct, then by including the effect of the *partial* drag on the case we have just considered, the difference between $E'G$ and EG should disappear.

By Fresnel's hypothesis, in the time l/c_b, the moving glass should have dragged the light over a distance of $lv/c_b(1-1/n^2)$. This should just be equal to the discrepancy between $E'G$ and EG or equal to $E'E$ in the diagram. Now

$$E'E = E'G - EG = \frac{lnv}{c} - \frac{lv}{nc} \tag{3.5}$$

and since $c_b = c/n$, if we divide the numerator and denominator of both fractions by n and factor we obtain

$$E'E = \frac{lv}{c_b}\left(1 - \frac{1}{n^2}\right), \tag{3.6}$$

which is the desired result, indicating that the Fresnel partial dragging coefficient accounts for what is actually observed.

The Fresnel convection coefficient was subsequently confirmed for light projected through water moving relative to the surface of the Earth by Fizeau (1851) and a similar experiment was repeated by Michelson and Morley (1886) with considerably increased precision, with the convection coefficient again being confirmed.

3. Stokes

We have seen that Fresnel noted in his letter to Arago that he could not see how to possibly account for stellar aberration on the basis of the wave theory if the Earth were assumed to completely drag the aether along with it, so that the velocity of the aether would be equal to the absolute velocity of the earth. In 1845 the British physicist G. G. Stokes published a short paper in the *Philosophical Magazine* which showed how this could be done. The Stokes theory of aberration was of some influence during the nineteenth century, at least until about 1886–7, for reasons that have to do in part with the confirmation of Fresnel's partial dragging coefficient for moving water by Michelson and Morley (1886), and in part with a criticism of Stokes' theory by Lorentz in 1886.

Until 1886, however, it seemed that Stokes' and Fresnel's theories were each adequate to account for aberration phenomena.

Stokes assumed that the Earth completely dragged the aether along with it in its orbit, but that it did so only near its surface. The velocity of the aether is, however, claimed to be identical at every point on the Earth's surface and apparently equal to the absolute velocity of the Earth in the universe. Out in space, however, "at no great distance" from the Earth, the aether was supposed to be in a state of absolute rest. With these ideas in mind, Stokes considered how aberration phenomena might be explained.

He began by noting that the direction of the wave front of the starlight impinging on the Earth should be dependent on both the velocity of the light through the aether and on the velocity of the aether streaming near the Earth. Stokes analyzed the possible effect of the aether's supposed motion on the equation of the wave front.[†]

Let u, v, w be the velocity of the aether stream in the neighborhood under consideration, i.e. somewhere above the Earth. Assuming that the axis z of an xyz coordinate system is in the direction of the propagation of a plane wave, the equation of the wave front is:

$$z = C + Vt + \zeta \tag{3.7}$$

where C is some arbitrary constant, V the velocity of light, and t the time, and ζ a small quantity and a function of x, y, and t. ζ will turn out to be a measure of the rotation or aberration of the wave front as caused by u, v, w.

Stokes confined himself to first-order quantities, dropping terms involving squares of the ratio of velocity of the Earth to the velocity of light. The direction cosines of a normal to the wave front are:

$$\cos \alpha = -\frac{d\zeta}{dx}, \quad \cos \beta = -\frac{d\zeta}{dy}, \quad \cos \gamma = 1. \tag{3.8}$$

[†] Stokes' analysis is presented in more elegant terms in Lorentz (1901).

At a distance $V\,dt$ along the normal, the coordinates will be altered from what they would be if there were no aether stream velocity. The coordinates, taking the moving aether into account, will be:

$$x' = x + \left(u - V\frac{d\zeta}{dx}\right) dt,$$
$$y' = y + \left(v - V\frac{d\zeta}{dy}\right) dt, \qquad (3.9)$$
$$z' = z + (w + V)\, dt.$$

Substituting $F(x, y, t)$ for ζ, and employing (3.7), expanding the resulting expression neglecting dt^2 and the square of the aberration constant, and solving for z, Stokes obtained:

$$z = C + Vt + \zeta + (w + V)\, dt. \qquad (3.10)$$

Using (3.7) again, this time computing the wave front's equation at time $t + dt$, Stokes got:

$$z = C + Vt + \zeta\left(V + \frac{d\zeta}{dt}\right) dt. \qquad (3.11)$$

Comparison term by term of (3.10) and (3.11) yielded:

$$\frac{d\zeta}{dt} = w \quad \text{or} \quad \zeta = \int w\, dt. \qquad (3.12)$$

But since ζ is small, $\int w\, dt$ may be approximately represented by $\int w\, dz/V$, the equation for the wavefront (3.7) becoming:

$$z = C + Vt + \frac{1}{V}\int w\, dz. \qquad (3.13)$$

Comparison of (3.13) with the equations for the direction cosines of the normal to the wavefront gives:

$$\alpha - \frac{\pi}{2} = \frac{1}{V}\int \frac{dw}{dx}\, dz, \quad \beta - \frac{\pi}{2} = \frac{1}{V}\int \frac{dw}{dy}\, dz. \qquad (3.14)$$

The terms under the integral sign are measures of the rotations of the wave front about the y and x axes. Integrated, the expressions represent the components of the total rotation of the wave front— i.e. the aberration—due to the aether stream velocity. The limits of the integration must range from the Earth's surface to a point out in space where the effect of the Earth's motion on the aether is imperceptible. The equations of (3.14) will account for the aberration that is actually observed if u, v, and w are such that $u\,dx + v\,dy + w\,dz$ is an exact differential. Physically this amounts to the assumption that the aether is irrotational: that it has no vortices in its stream. If this is so, then:

$$\frac{dw}{dx} = \frac{du}{dz}, \quad \frac{dw}{dy} = \frac{dv}{dz}$$

and substitution of this in (3.14) yields:

$$\alpha_2 - \alpha_1 = \frac{1}{V} \int_1^2 \frac{du}{dz} \cdot dz = \frac{u_2 - u_1}{V},$$
$$\beta_2 - \beta_1 = \frac{1}{V} \int_1^2 \frac{dv}{dz} \cdot dz = \frac{v_2 - v_1}{V}.$$
(3.15)

Stokes applied equations (3.15) to a star. Point 1 is sufficiently distant so that $u_1 = 0$ and $v_1 = 0$. The plane xz was chosen so that it passed through the direction of the Earth's motion. Then v_2 equalled 0, and $\beta_2 - \beta_1$ also equalled zero. Consequently:

$$\alpha_2 - \alpha_1 = \frac{u_2}{V}$$

which is the aberration constant, and Bradley's well-known result.

4. Michelson and Morley

In 1881 when A. A. Michelson first performed his interferometer experiment Fresnel's explanation of aberration was generally accepted.[†] If, as in Fresnel's theory, the aether was indeed stationary with the Earth moving through it, the time it would take for a light wave to pass between two points on the surface of the Earth would be different if it were moving in the direction of the Earth's motion, or opposite to this motion. Because of cancellation effects involved in passing the light to and fro over the same path, the effect of the Earth's motion is extremely small, of the second order of v/c, or about one part in 10^8. Nevertheless, Michelson discovered a means of measuring this quantity.

He constructed an apparatus known as an interferometer, which permitted two rays of light which traveled over paths at right

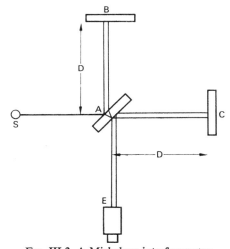

FIG. III.3. A Michelson interferometer.

angles to one another to recombine and interfere. The original interferometer is diagrammed in Fig. III.3. The light from a lamp or a sodium flame positioned at S is divided by the partially silvered

[†] For a more complete exposition of Michelson's work see R. Shankland's (1964) excellent article on the interferometer experiment. Lloyd Swenson's (1962) dissertation is also worth reading in this connection.

mirror located at A and moves to B and C, from which it is reflected and then recombined at A. If the paths AB and AC are equal, the two rays interfere along AE. The interference shows up in the eyepiece positioned at E as thin dark fringes or bands in the white or yellow light. Ordinarily monochromatic light is used for alignment and then white light can be substituted and colored fringes sought for. In the latter case the fringes disappear very easily and can be used as a careful check on the equality of the paths.

If we assume with Fresnel that the Earth moves through the aether without dragging it along, the amount that the fringes of the interferometer should shift when the interferometer is turned through an angle of 90° can be computed as follows: We let c be

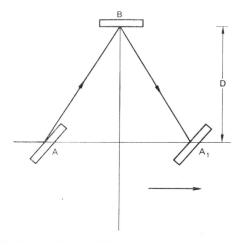

FIG. III.4. The "vertical" path of light in the interferometer as considered from the point of view of Fresnel's aether theory.

the velocity of light, v the velocity of the earth through the aether, D the distance AB or AC, T the time light requires to pass from A to C, and T_1 the time required for the light to return from C to A_1. The distance AA_1 is shown in Fig. III.4 and is due to the movement of the interferometer during the time required for the passage of the light from the partly silvered mirror to the reflecting mirrors and back again. In both cases, however, the horizontal distance traversed is D, with the velocity of the light in the first case being $c-v$, as it "bucks" the aether wind, and $c+v$ on its return.

Accordingly we have:

$$T = \frac{D}{(c-v)}, \quad T_1 = \frac{D}{(c+v)}.$$

The total time required for transit is then:

$$T+T_1 = \frac{D}{(c-v)} + \frac{D}{(c+v)} = \frac{2Dc}{c^2-v^2}.$$

The total distance traversed by the light, then, is:

$$\frac{2Dc^2}{(c^2-v^2)} = 2D\left(1+\frac{v^2}{c^2}\right)$$

dropping terms of the fourth and higher order of v/c in the expansion.

The length of the "vertical" path can be computed in several ways. In his 1881 analysis, Michelson overlooked the fact that the vertical path was actually a triangular path. In 1887 Michelson and Morley used for the value of the "vertical" path distance the expression:

$$2D\sqrt{\left(1+\frac{v^2}{c^2}\right)} \quad \text{or expanding,} \quad 2D\left(1+\frac{v^2}{2c^2}\right)$$

if we neglect powers higher than the second order of v/c. This first expression differs from the expression usually employed today for the distance, which is

$$\frac{2D}{\sqrt{(1-v^2/c^2)}}.$$

The difference vanishes, however, if we restrict ourselves to second order quantities, and is most likely due to an approximation involved in the calculation of the "vertical" path length.[†]

The difference in path then, in the 1887 analysis, is given by the quantity $D(v^2/c^2)$ which is obtained by subtracting the two path length calculations.

If the 1881 "vertical" path length distance is used, however, the path difference is equal to $2Dv^2/c^2$. The error in Michelson's calculations was pointed out to him soon after he published the first results, and a detailed examination of the experiment was published somewhat later by H. A. Lorentz (1886) as part of a long paper on aberration.

Michelson's 1881 experiment was performed in Potsdam in April of that year, and Michelson's computation of the expected aether drift took into account the direction of the Earth's motion at that time of the year. His reasoning told him that the fringe displacement due to the aether wind should be about 0.1 of a fringe, maximum, as the interferometer was rotated. The calculated fringe shift is plotted in Fig. III.5, as a dotted line. The solid line is the observed shift. Michelson concluded his 1881 paper with the following comment, which is all the more interesting because of the reference to Stokes' theory and the quoting of Stokes' views on aberration theories:

> The interpretation of these results is that there is no displacement of the interference bands. The result of the hypothesis of the stationary aether is thus shown to be incorrect, and the necessary conclusion follows that the hypothesis is erroneous.

[†] The difference between the contemporary expression for path length and Michelson and Morley's expression is easily seen if we write the contemporary expression as

$$2D \sqrt{\left(1 + \frac{v^2}{c^2} \cdot \frac{1}{1 - v^2/c^2}\right)}$$

to which it is exactly equal. Michelson and Morley apparently disregarded this small factor in their derivation of the vertical path length.

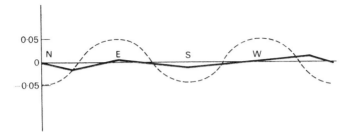

FIG. III.5. Graphical representation of A. A. Michelson's anticipated results (dotted line) and his experimental results (solid line). (After Michelson, 1881.) The ordinate represents the amount of fringe shift, and the abscissa the compass direction of one axis of the interferometer.

This conclusion directly contradicts the explanation of the phenomenon of aberration which has been hitherto generally accepted, and which presupposes that the Earth moves through the aether, the latter remaining at rest.

It may not be out of place to add an extract from an article published in the *Philosophical Magazine* by Stokes in 1846.

"All these results would follow immediately from the theory of aberration which I proposed in the July number of this magazine [this is the theory discussed under section 3 above—K. F. S.]: nor have I been able to obtain any result admitting of being compared with experiment, which would be different according to which theory we adopted. This affords a curious instance of two totally different theories running parallel to each other in the explanation of phenomena. I do not suppose that many would be disposed to maintain Fresnel's theory, when it is shown that it may be dispensed with, inasmuch as we would not be disposed to believe, without good evidence, that the ether moves quite freely through the solid mass of the Earth. Still it would have been satisfactory, if it had been possible to have put the two theories to the test of some decisive experiment."

In his article on aberration, Lorentz (1886) not only criticized Michelson's calculations by pointing out the missing factor of 2, but he also argued that if the correct values were used, that the experimental error involved would be enough to call into doubt any rejection of Fresnel's theory. Lorentz had a very specific reason for wishing to defend a version of Fresnel's theory against experimental refutation, for he had in the same article shown that Stokes' theory assumed boundary conditions which were inconsistent with its theoretical assumptions. Specifically, Stokes required his aether to be irrotational, that is, it had to have a velocity potential. But Stokes also assumed that the velocity of the aether everywhere on the surface of the Earth was the same, and Lorentz was able to show that this condition is inconsistent with the assumption of a velocity potential. About a dozen years later, Max Planck (see Lorentz 1899b) briefly resuscitated Stokes' aether by showing that Stokes' two assumptions could be made consistent if the aether were as compressible as a gas that follows Boyle's law, and if, accordingly, its density was great near the surface of the Earth, and smaller as the distance from the Earth increased. Such an increase in aether density could itself be accounted for if the aether were attracted by the Earth's gravity. Planck's hypothesis had no other observable consequences, however, and as the Fresnel aether had in a sense been very successfully absorbed into Lorentz' electron theory in 1892, Planck's notions did not attract very much attention. Lorentz (1899b), however, commented on them and criticized them.

In his 1886 essay Lorentz had unequivocally sided with the Fresnel aether, though he analyzed it in somewhat more sophisticated terms than Fresnel had done, ascribing a velocity potential to it that would yield the partial dragging coefficient within ponderable bodies, but assuming that the aether was stationary in empty space. Such an aether implies nearly the same positive interferometer results that Michelson had anticipated (except for the factor of 2), and Lorentz did not deny that the interferometer experiment would not have a positive result were it performed again with more precision.

At the urging of Lord Rayleigh, Michelson repeated his experiment again in 1887 with the assistance of his colleague at Case Institute, E. W. Morley. This time the calculations were corrected for the influence of the Earth's motion on the "vertical" ray. The precision of the experiment was also increased by lengthening the path by using four mirrors at the extreme reflecting points rather than one. A diagram of the improved interferometer appears in the appended Michelson–Morley selection and need not be reproduced here. The observations were made in July of 1887 at different hours. This time only the Earth's orbital velocity figured in the calculations, and the predicted fringe displacement was, with the increased path, computed to be about 0·4 of a fringe. The curves of one-eighth of the predicted displacement and the displacement observed are given in Fig. 6 of the Michelson–Morley selection. The observed displacement was somewhere between one-twentieth and one-fortieth of the predicted value. Michelson and Morley's assessment of their result was as follows:

> It appears from all that precedes reasonably certain that if there be any relative motion between the earth and the luminiferous aether, it must be small; quite small enough entirely to refute Fresnel's explanation of aberration. Stokes has given a theory of aberration which assumes the aether at the earth's surface to be at rest with regard to the aether, and only requires in addition that the relative velocity have a potential; but Lorentz shows these conditions are incompatible. Lorentz then proposes a modification which combines some ideas of Stokes and Fresnel, and assumes the existence of a potential, together with Fresnel's coefficient. If now it were legitimate to conclude from the present work that the aether is at rest with regard to the earth's surface, according to Lorentz then there could not be a velocity potential, and his own theory also fails.

In Chapter VI I shall discuss Lorentz' response to this refutation, and also touch on Einstein's "explanation" of the null result of the interferometer experiment.

CHAPTER IV

THE ELASTIC SOLID AETHER

As I pointed out in Chapter I, nineteenth-century aether theories were largely attempts to formulate explanations of optical, and later, electromagnetic phenomena, in mechanical terms. In these theories some law or principle of mechanics was asserted, from which, subject to the proper boundary and symmetry conditions, laws which possessed a formal analogy with optical and electromagnetic laws were derived. This tack was not universally taken, and Lorentz' post 1892 work does constitute a significant exception. Nevertheless, in order to understand nineteenth-century aether theory in many of its aspects, a rudimentary knowledge of the theoretical mechanics of the period is essential. The most important mechanical approaches during this time are the variational formulations of Lagrangian analytical mechanics and, later in the century, the analyses presented in terms of Hamilton's principle or the "principle of least action".

1. Introduction to Nineteenth-century Mechanics

In 1788 in his *Mechanique Analytique*, Lagrange presented an account of mechanics which eliminated the dependence of the subject on the Newtonian geometrical reasoning. The name "analytical mechanics" has been appended to Lagrange's approach because his account was algebraic or "analytical" rather than geometrical or "synthetic". Newton's mechanics was certainly of the latter character, and many of Newton's modifiers, such as D'Alembert, the Bernoullis, and, to some extent, even Euler, utilized geometrical reasoning. For more detailed discussion on this point, the reader should refer to Ernst Mach's *The Science of Mechanics*, pp. 560–2.

Lagrangian mechanics is logically equivalent to Newtonian mechanics, though it does represent an advance with respect to mathematical elegance. Furthermore, its approach is extremely general and quite powerful for analyzing many mechanical problems which would be cumbersome and difficult to solve in the Newtonian formulation. Similar advantages accrue to the still later analysis of mechanics by Hamilton, who in 1834 developed the science on the basis of a principle later known variously as "Hamilton's principle" or the "principle of least action", though these two notions are considered somewhat different today.

It is not very difficult to state and derive the relationships between the Lagrangian approach, the Hamilton principle, and today's least action principle. (Though the mathematically illiterate may skip the following section, they do so unadvisedly, as Lagrangian and Hamiltonian methods are extensively used by Green, MacCullagh, Fitzgerald, and Larmor in their aether theories.)

Lagrangian dynamics develops from a principle of statics known as d'Alembert's principle of virtual velocities, or better, virtual (or arbitrary) "displacements". If we have some interconnected system with various internal forces acting on the parts, if the system is to be in equilibrium, the sum of all the forces resolved into components in the x, y, and z axis directions, each multiplied by an infinitely small displacement δx, δy, δz in those directions, must add up to zero. The interconnections within the system will establish constraints or relations between the infinitesimal displacements. This principle can be expressed in the formula:

$$\Sigma(F_x \, \delta x + F_y \, \delta y + F_z \, \delta z) = 0. \tag{4.1}$$

This principle is extended to systems in motion in the following way. Consider the system to be composed of mass points m_1, m_2, ... m_n, and refer the system to a Cartesian coordinate system of mutually perpendicular axes x, y, z, as in the static case. Let resolved forces act on each mass point, their values being represented by X_1, Y_1, Z_1, for m_1, X_2, Y_2, Z_2, for m_2, etc. Assume such forces produce virtual displacements δx_1, δy_1, δz_1, and δx_2, δy_2, δz_2, etc. Thus far nothing new has been added to the above static case except a change of symbolism. In fact equation (4.1) in this new symbolism would be:

$$\sum_{i=1}^{n} (X_i \, \delta x_i + Y_i \, \delta y_i + Z_i \, \delta z_i) = 0. \tag{4.2}$$

Now consider the forces X, Y, Z acting on each mass point as *impressed forces* which produce motions within the interconnected system such that particle m_1 moves with acceleration components

$$\frac{d^2x_1}{dt^2}, \quad \frac{d^2y_1}{dt^2}, \quad \frac{d^2z_1}{dt^2}.$$

If each of these acceleration terms is multiplied by m_1, the product is equivalent to the net force acting on m_1. These net forces are termed *effective forces*. They are not necessarily equivalent to the *impressed forces* cited above. The extension of d'Alembert's statics principle to dynamics then is made as follows: the difference between the impressed force components and the effective force components must be such that the sum of them (the differences) adds up to zero, as they produce no motion. In equation form this statement amounts to:

$$\sum_{i=1}^{n} \left[\left(X_i - m_i \frac{d^2x_i}{dt^2} \right) \delta x + \left(Y_i - m_i \frac{d^2y_i}{dt^2} \right) \delta y_i + \left(Z_i - m_i \frac{d^2z_i}{dt^2} \right) \delta z \right] = 0. \tag{4.3}$$

This can also be stated in more compact vector notation as:

$$\sum_{i=1}^{n} \left(F_i - m_i \frac{d^2r}{dt^2} \right) \cdot \delta r = 0 \tag{4.4}$$

where r is a displacement vector and F is a force vector. Finally, the principle can be stated in a form in which it is easily applicable to continuum mechanics, such as we shall find in the various nineteenth-century aether theories, by replacing the particles' masses by a density function multiplied by a differential volume, transposing the force function to the right hand side, and assuming that it acts on a volume element. We then obtain in an integral formulation:

$$\iiint \varrho \left(\frac{d^2x}{dt^2}\,\delta x + \frac{d^2y}{dt^2}\,\delta y + \frac{d^2z}{dt^2}\,dz \right) dx\,dy\,dz = \iiint F\,dx\,dy\,dz. \tag{4.5}$$

Equations (4.3), (4.4), and (4.5) are essentially equivalent to one another. Because of its compactness of form, I shall use (4.4) to show how Hamilton's principle can be obtained from the Lagrange formulation.[†]

I shall restate (4.4) using dot notation for differentiation with respect to time as:

$$\Sigma(\boldsymbol{F}_i - m_i\ddot{\boldsymbol{r}}_i)\cdot\delta\boldsymbol{r}_i = 0. \tag{4.4}$$

Now from the properties of δ it follows that:

$$\ddot{\boldsymbol{r}}_i\cdot\delta\boldsymbol{r}_i = \frac{d}{dt}(\dot{\boldsymbol{r}}_i\cdot\delta\boldsymbol{r}_i) - \frac{1}{2}\delta(v_i)^2,$$

where $v_i^2 = \dot{\boldsymbol{r}}_i\cdot\dot{\boldsymbol{r}}_i$, v being interpreted as velocity. Consequently:

$$\frac{d}{dt}\Sigma m_i\dot{\boldsymbol{r}}_i\cdot\delta\boldsymbol{r}_i = \delta\Sigma\frac{1}{2}m_iv_i^2 + \Sigma m_i\ddot{\boldsymbol{r}}_i\cdot\delta\boldsymbol{r}_i \tag{4.6}$$

in which the summation runs over i from 1 to n. Integrating both sides of (4.6) from $t = t_0$ to $t = t_1$, we obtain, setting $T = \Sigma\frac{1}{2}m_iv_i^2$:

$$\Sigma m_i\dot{\boldsymbol{r}}_i\cdot\delta\boldsymbol{r}_i\bigg]_{t_0}^{t_1} = \int_{t_0}^{t_1}(\delta T + \Sigma m_i\ddot{\boldsymbol{r}}_i\cdot\delta\boldsymbol{r}_i)\,dt. \tag{4.7}$$

It can now be assumed that the systems to be analyzed are ones whose initial and final positions—i.e. at t_0 and t_1—are the same.

[†] This derivation essentially follows Lindsay and Margenau (1957), pp. 131–2.

It follows from this restriction that r_i at t_0 and t_1 is equal to zero. But then the left side of (4.7) is zero and we obtain:

$$\int_{t_0}^{t_1} (\delta T + \Sigma m_i \ddot{\boldsymbol{r}}_i \cdot \delta \boldsymbol{r}_i)\, dt = 0. \tag{4.8}$$

Suppose now, and this will be important for Green's and MacCullagh's aether investigations later, that there exists a function V of the rectangular coordinates of the parts of the system such that:

$$\Sigma \boldsymbol{F}_i \cdot \boldsymbol{r}_i = \Sigma m_i \ddot{\boldsymbol{r}}_i \cdot \delta \boldsymbol{r}_i = -\delta V. \tag{4.9}$$

Then (4.8) becomes:

$$\delta \int_{t_0}^{t_1} (T-V)\, dt = 0. \tag{4.10}$$

This is "Hamilton's principle", but it is often referred to by nineteenth-century physicists as the "Principle of Least Action". In words it can be stated as "Assuming a conservative system, the system changes in such a way as to minimize (over short intervals of time) the action integral". In contemporary works, the "principle of least action" is usually understood as the assertion that:

$$\delta \int_{t_0}^{t_1} 2T\, dt = 0 \tag{4.11}$$

which is closer to Maupertuis' principle of least action. Equation (4.11) is less general than (4.10) since (4.11) is restricted to cases in which, during changes in the system, the total energy $U = T+V$, is constant and the same over every varied path. In (4.10), however, the variations of the paths are perfectly general except at the end points.

Equation (4.11) can be stated still another way by making use of the total energy equation mentioned immediately above, and by introducing an element of arc length $ds = (2T)^{\frac{1}{2}} dt$. We then obtain:

$$\delta \int_{t_0}^{t_1} (U-V)^{\frac{1}{2}}\, ds = 0 \tag{4.12}$$

which is still another way of stating the "Principle of Least Action".

The rationale for the digression into these variational formulations of mechanics is to prepare the reader for the nineteenth-century inquiries into the aether which were conducted by Green and MacCullagh, as regards the optical aether, and by Maxwell, Fitzgerald, and Larmor, as regards the electromagnetic aether. For the latter two, at least, MacCullagh's optical aether was Maxwell's electromagnetic aether. I shall discuss the contributions of Green and MacCullagh in this chapter, and consider the others' work in the following chapter.

2. The Development of the Elastic Solid Theory of the Aether

As was discussed earlier in Chapter II, the Young and Fresnel theories of the aether were not dynamical theories in any real sense. At the time they were developed, the proper equations and solutions describing wave motion in an elastic solid were not available. It was left to the French mathematicians, Navier, Cauchy, and Poisson to develop a mathematical theory of vibrations in a mechanical elastic solid, and to Cauchy to first apply these analyses to the wave motion of light.[†] Cauchy's contributions are of considerable importance in the elastic solid theory of light and his work was highly valued by his successors in this field. Cauchy first presented his molecular theory of the aether in 1830, and later in 1836 and 1839 presented two more somewhat different theories of the optical aether. Because of the brief, and therefore necessarily eclectic character of this book, however, I can do no more than cite Cauchy's contributions, and must refer the interested reader to other sources.

For various reasons, I have decided to include the full text of a paper by the English physicist and mathematician George Green.

[†] See Whittaker (1960), I, pp. 128–33, for a good discussion of the French mathematicians.

Green's theory of the aether was first presented in 1837, and came to have considerable influence on the development of the elastic solid theory through the enthusiastic missionary work of William Thomson, later Lord Kelvin. The virtue of Green's theory, aside from its historical influence, lies in its simplicity, its generality, its adaptability to change, e.g. by Lord Rayleigh (1871) and Lord Kelvin (1888), and in its physical naturalness. It constitutes a particularly elegant example of a style of argument and a type of theory which was of considerable influence a little more than 100 years ago, but which is almost completely forgotten today. Some of Green's other work has enjoyed a better fate, and Green's contributions to function theory, potential theory, and electrical theory are fairly well known. "Green's functions" are also extensively used in contemporary differential and integral equation theory.

3. Green's Aether Theory

Green's approach to the elastic solid theory of the aether is through Lagrangian mechanics applied to matter in bulk. Though Green, like Cauchy, supposes a molecular structure for the aether, his analysis is sufficiently independent of this structure so as also to be able to characterize a continuous aether.

Green begins by pointing out that Cauchy's theory (apparently he is referring only to Cauchy's "first theory") involves an assumption of forces acting between aether particles in which the direction of the action of forces is always along a line joining any two particles. This assumption or principle of central forces, common among the Newtonian-influenced French physicists of the nineteenth century, seemed "rather restrictive" to Green. The assumption which Green wished to substitute in its place was a version of d'Alembert's principle as developed in Lagrange's mechanics, about which I have spoken above. Green wrote:

> The principle selected as the basis of the reasoning contained in the following pages is this: In whatever way the elements of any material

system may act upon each other, if all the internal forces exerted be multiplied by the elements of their respective directions the total sum for any assigned portion of the mass will always be the exact differential of some function.

Green then turned his attention to a difficulty which was to appear again and again in elastic solid theories. In elastic media, with some peculiar and quite questionable exceptions to be touched on in the later sections of this chapter on MacCullagh's and Kelvin's aethers, a disturbance produces *two* spherical waves: one is a longitudinal compressional wave, the other a transverse wave. One of the major problems that was faced by all elastic solid theorists was to eliminate the longitudinal wave, or at least to eliminate its observable consequences, for experiments, such as the Arago–Fresnel experiment cited in the previous chapter, indicated that light was a purely transverse wave. We saw how Fresnel eliminated it in his quasi-mechanical theory—simply by hypothesis. Cauchy, in his first theory, did not seek to eliminate it, as he thought it actually existed and might be degraded as heat, and that experiment would disclose its effects. This, however, was not the case.

Green's manner of eliminating the longitudinal wave is to conceive of his elastic solid aether as so rigid—in the sense of being resistant to compression—that the velocity of the longitudinal wave becomes practically equal to infinity. This resistance to compression, though, is a relative resistance as we shall see below. In the introductory section of his paper he anticipates the results of his inquiry into the aether, and tells us that the solution of the equation of motion of his medium will contain two arbitrary coefficients, A and B, whose values depend on the unknown internal constitution of the aether. The velocity of the longitudinal wave is proportional to $\sqrt{A}$ and the velocity of the transverse wave to $\sqrt{B}$. The effect of the longitudinal wave must be eliminated since in Green's theory, even if it itself be incapable of affecting the eye, it will give rise to a new transverse wave at a reflecting–refracting surface unless the ratio A/B equals zero or infinity, which would be visible. Green argues that if A/B is less than $\frac{4}{3}$, the medium is unstable. Consequently the velocity of the longitudinal wave must be very great compared with the velocity of ordinary light, i.e. approximately infinite. (If A/B were less than $\frac{4}{3}$, an increase in pressure would produce an increase in volume, and the medium might possibly explode.) The constant A is, roughly, a measure of the aether's resistance to compression or change of volume and the constant B a measure of the aether's resistance to distortion with no change in volume, e.g. to twist. Consequently the ratio of A to B then, though practically infinite, is really a relative *ratio* of the resistance against compression to the resistance against distortion. Accordingly the aether need not be any more resistant to compression than the rarest known gas, if the resistance to distortion is exceedingly small. Even though the aether must support transverse waves moving at a velocity approximately equal to 3×10^8 meters/second, such a small compressibility is not ruled out on the foregoing suppositions if we can assume that the aether *density* is exceedingly small. Accordingly, Green's assumptions regarding the nature of the aether were not necessarily inconsistent with what was known about the Universe when he wrote, and the planets could move through such an aether.

For the constitution of the aether Green supposed, as many had before him, that it consisted of a large number of very small aether particles interacting with one another via very short-range molecular forces. Let x, y, and z be the equilibrium coordinates of any particle, and let $x+u$, $y+v$, and $z+w$ be the coordinates of the same particle in a state of motion. In accordance with Green's version of d'Alembert's principle we may write:

$$\Sigma \left\{ Dm \frac{d^2u}{dt^2} \delta u + Dm \frac{d^2v}{dt^2} \delta v + Dm \frac{d^2w}{dt^2} \delta w \right\} = \Sigma \, \delta\phi DV \quad (4.13)$$

where $Dm(d^2u/dt^2)$, for example, is equivalent to the internal force exerted in the x direction, δu is the x component of the virtual

THE ELASTIC SOLID AETHER 49

displacement of the particle, and $\delta\phi$ is the variation of the function sought for. The summed product $\Sigma\,\delta\phi DV$ represents the work given out by the differential volume DV in passing from an equilibrium state to the new non-equilibrium state. Dm is not the mass of a single aether particle, but is, rather, the mass of a very small differential volume which nevertheless contains a large number of aether molecules.

The equation (4.13) can be put into integral form so as to closely resemble in form equation (4.5) which was discussed above in the introduction to nineteenth-century mechanics. Green in fact does let (4.13) pass into the integral form by introducing an aether density term, ϱ, and rewriting (4.13) as:

$$\int\int\int \varrho\,dx\,dy\,dz\left\{\frac{d^2u}{dt^2}\,\delta u+\frac{d^2v}{dt^2}\,\delta v+\frac{d^2w}{dt^2}\,\delta w\right\} \\ =\int\int\int dx\,dy\,dz\,\delta\phi. \tag{4.14}$$

Green argues that ϕ is a function entirely dependent on the internal actions of the particles of the medium on each other, and accordingly is a function of the compression and distortion of the medium. $\delta\phi$ must be an exact differential for Green because if the converse held true perpetual motion would be possible. Green wrote before conservation of energy (other than conservation of *vis viva*) was accepted, and he seems to base his argument of the form of $\delta\phi$ on the conservation of *work*.

To obtain the form of ϕ Green considers the effects of an arbitrary distortion administered to the differential element of volume. He lets dx, dy, and dz represent the sides of a rectangular differential element, and dx', dy', and dz' the element in a state of distortion (and compression). Green introduces small quantities s_1, s_2, and s_3 to represent elongations and α, β, and γ to represent angular distortions or principal shearing strains. These quantities are related to the sides of the differential elements by:

$$dx' = dx(1+s_1) \quad dy' = dy(1+s_2) \quad dz' = dz(1+s_3) \quad (4.15)$$

$$\alpha = \cos\left\langle\begin{array}{l}dy'\\dz'\end{array}\right. \quad \beta = \cos\left\langle\begin{array}{l}dx'\\dz'\end{array}\right. \quad \gamma = \cos\left\langle\begin{array}{l}dx'\\dy'\end{array}\right. \quad (4.16)$$

The notation $\cos\left\langle\begin{array}{l}dy'\\dz'\end{array}\right.$ indicates the cosine of the angle formed by the line elements dy' and dz'. Green later shows that if we neglect higher-order quantities, these small quantities can be defined in terms of the motion of the aether particle as:

$$\begin{array}{ccc} s_1 = \dfrac{du}{dx} & s_2 = \dfrac{dv}{dy} & s_3 = \dfrac{dw}{dz} \\[2mm] \alpha = \dfrac{dw}{dy}+\dfrac{dv}{dz} & \beta = \dfrac{dw}{dx}+\dfrac{dv}{dz} & \gamma = \dfrac{du}{dy}+\dfrac{dv}{dx} \end{array} \quad (4.17)$$

The important function ϕ then is considered to be a function of these six quantities or:

$$\phi = \text{function } (s_1, s_2, s_3, \alpha, \beta, \gamma). \quad (4.18)$$

The determinate form of ϕ is obtained by a complicated series of steps:

(1) First ϕ is expanded into a series:

$$\phi = \phi_0 + \phi_1 + \phi_2 + \phi_3 + \ldots \quad (4.19)$$

each ϕ_i being an ith degree function of $s_1, s_2, s_3, \alpha, \beta, \gamma \ldots$.

(2) Certain plausible boundary conditions are then imposed. Since $\phi_0 =$ a constant, $\delta\phi_0 = 0$. By hypothesis, at equilibrium $u = 0$, $v = 0$, and $w = 0$—i.e. the medium is unstrained at equilibrium position—so it follows that $\iiint dx\,dy\,dz\,\delta\phi_1 = 0$. (If the medium were under a pressure at equilibrium, ϕ_1 would be a func-

tion with six arbitrary constants.) ϕ_3 and ϕ_4, etc., are considered to be exceedingly small with respect to ϕ_2, so that the general function ϕ reduces to ϕ_2, which, since it is a homogeneous function of six independent variables of the second order, contains no more than twenty-one arbitrary constants in its most general form.

(3) Green then assumes that the medium is unlike a crystalline body, that is, that it is symmetrical with respect to three rectangular planes. If this is so, the twenty-one arbitrary constants reduce to nine, and we have:

$$\phi_2 = G\left(\frac{du}{dx}\right)^2 + H\left(\frac{dv}{dy}\right)^2 + I\left(\frac{dw}{dz}\right)^2 + L\alpha^2 + M\beta^2 + N\gamma^2 \\ + 2P\frac{dv}{dy} \cdot \frac{dw}{dz} + 2Q\frac{dv}{dx} \cdot \frac{dw}{dz} + 2R\frac{du}{dx} \cdot \frac{dv}{dy} \quad (4.20)$$

where G, H, I, L, M, N, P, Q, and R are the nine arbitrary constants.

(4) If, furthermore, ϕ_2 is restricted to a medium with symmetry around one axis, Green shows that:

$$\begin{aligned} G &= H = 2N + R, \\ L &= M, \\ P &= Q. \end{aligned} \quad (4.21)$$

And, finally, if ϕ_2 is symmetrical with respect to all three mutually perpendicular Cartesian axes, that is, if the medium is isotropic, we get:

$$\begin{aligned} G &= H = I = 2N + R, \\ L &= M = N, \\ P &= Q = R. \end{aligned} \quad (4.22)$$

The equation (4.20) can accordingly be simplified by utilizing these relations among the constants. Introducing two more constants, apparently for aesthetic purposes, $A = 2G$ and $B = 2L$, Green obtains his determinate form of ϕ_2:

$$2\phi_2 = -A\left(\frac{du}{dx}+\frac{dv}{dy}+\frac{dw}{dz}\right)^2 - B\left\{\left(\frac{du}{dy}+\frac{dv}{dx}\right)^2 \right.$$
$$+ \left(\frac{du}{dz}+\frac{dw}{dx}\right)^2 + \left(\frac{dv}{dz}+\frac{dw}{dy}\right)^2 \quad (4.23)$$
$$\left. -4\left(\frac{dv}{dy}\cdot\frac{dw}{dz}+\frac{du}{dx}\cdot\frac{dw}{dz}+\frac{du}{dx}\cdot\frac{dv}{dy}\right)\right\}.$$

This is the general form of what we would now call the potential energy function of the aether in a non-crystalline medium.

The general equation of motion, (4.14) above, is then written for an aether disturbance moving from one substance into another across a surface, which will be the reflecting–refracting surface, and which is taken to be an infinite horizontal plane. If the aether density in the upper medium is ϱ and the density of the lower medium ϱ_1, and ϕ_2 and $\phi_2^{(1)}$ the respective work (or potential energy) functions, by (4.14) we may write two equations and sum them to obtain:

$$\iiint \varrho \, dx \, dy \, dz \left\{\frac{d^2u}{dt^2}\delta u + \frac{d^2v}{dt^2}\delta v + \frac{d^2w}{dt^2}\delta w\right\}$$
$$+ \iiint \varrho_1 \, dx \, dy \, dz \left\{\frac{d^2u_1}{dt^2}\delta u_1 + \frac{d^2v_1}{dt^2}\delta v_1 + \frac{d^2w_1}{dt^2}\delta w_1\right\} \quad (4.24)$$
$$= \iiint dx \, dy \, dz \, \delta\phi_2 + \iiint dx \, dy \, du \, \delta\phi_2^{(1)}$$

in which the subscripts 1 distinguish quantities in the lower media. The integration for the triple integrals extends over the whole volume of the respective aethers.

Green then uses (4.23) to substitute the determinate form of ϕ_2 into (4.24) for both media, adding subscripts where necessary to distinguish quantities in the upper medium from their counterparts in the lower. Carrying out an integration by parts yields two complicated expressions: one a summed volume integral and another a surface integral. Both of these must equate separately to zero: the triple integral yielding the equations of motion of the medium, the surface integral giving the boundary conditions which hold at the interface between the two media (in this case where $x = 0$). The boundary conditions are obtained by the substitution of an additional requirement of continuity of the aether displacement into them. The equations of continuity are, obviously, $u = u_1$, $v = v_1$, $w = w_1$, from which one also directly obtains:

$$\delta u = \delta u_1, \quad \delta v = \delta v_1, \quad \text{and} \quad \delta w = \delta w_1.$$

Green obtains his equation of motion for the aether disturbance in the form of:

$$\varrho \frac{d^2 u}{dt^2} = A \frac{d}{dx} \left(\frac{du}{dx} + \frac{dv}{dy} + \frac{dw}{dz} \right)$$
$$+ B \left\{ \frac{d^2 u}{dy^2} + \frac{d^2 u}{dz^2} - \frac{d}{dx} \left(\frac{dv}{dy} + \frac{dw}{dz} \right) \right\}. \qquad (4.25)$$

There are, of course, three equations for each medium, and (4.25) is only one of one set. The equations of motion can be put in a more elegant form by rewriting (4.25), for example, as:

$$\varrho \frac{d^2 u}{dt^2} = (A - B) \frac{d}{dx} \left(\frac{du}{dx} + \frac{dv}{dy} + \frac{dw}{dz} \right) + B \nabla^2 u. \qquad (4.25')$$

It is perhaps easier to see the wave equation present in this form of the equation of motion.

The boundary conditions determined by the theory follow readily from the surface integrals and the conditions of continuity of aether displacement, and when $x = 0$, as at the interface chosen, become:

$$A\left(\frac{du}{dx}+\frac{dv}{dy}+\frac{dw}{dz}\right)-2B\left(\frac{dv}{dy}+\frac{dw}{dz}\right)$$

$$= A_1\left(\frac{du_1}{dx}+\frac{dv_1}{dy}+\frac{dw_1}{dz}\right)-2B_1\left(\frac{dv_1}{dy}+\frac{dw_1}{dz}\right)$$

$$B\left(\frac{du}{dy}+\frac{dv}{dx}\right) = B_1\left(\frac{du_1}{dy}+\frac{dv_1}{dx}\right)$$

$$B\left(\frac{du}{dz}+\frac{dw}{dx}\right) = B_1\left(\frac{du_1}{dz}+\frac{dw_1}{dx}\right) \quad (4.26)$$

Having obtained his equations of motion and his boundary conditions, Green turned his attention to attempting to derive Fresnel's sine and tangent laws for plane polarized waves, and to account for other optical phenomena, such as phase reversal on reflection. He considered the two cases of the plane polarized wave incident on his infinite horizontal plane: one in which the polarization is in the plane of incidence, the second in which the polarization is perpendicular to the plane of incidence. I shall only consider the first case in detail, though I shall comment on Green's second case.

The z axis is now chosen parallel to the intersection of the line formed by the intersection of the plane of the incident light and the interface. Polarization in this first case amounts to setting $u = 0$, $v = 0$, and $u_1 = 0$, $v_1 = 0$, as the vibrations now occur only in the z direction. The equation of motion, based on (4.25), thus reduces to:

$$\varrho\,\frac{d^2w}{dt^2} = B\left\{\frac{d^2w}{dx^2}+\frac{d^2w}{dy^2}\right\} \quad (4.27)$$

and introducing a new constant $\gamma^2 = B/\varrho$ Green obtains:

$$\frac{d^2w}{dt^2} = \gamma^2\left(\frac{d^2w}{dx^2}+\frac{d^2w}{dy^2}\right) \quad (4.28)$$

together with a similar equation for the lower medium, the only difference being in the addition of subscripts to w, γ, and B.

The boundary conditions of continuity and of (4.26) require in this case that:

$$w = w_1$$
$$B\frac{dw}{dx} = B_1 \frac{dw_1}{dx}. \tag{4.29}$$

In order to proceed further Green makes use of an additional hypothesis concerning the B quantities. He based his argument on the fact that the quantity A, which represents the compressibility, and on which the velocity of *longitudinal* waves depends, is independent of the nature of a *gas* as in sound wave theory. On the basis of this fact Green supposed that the B's in his media are also equal. It is clear that this is somewhat hypothetical, though it does agree with Young and Fresnel's views about the nature of the aether in different bodies in which refraction and wave velocity only depend on variations in aether density and not in rigidity.

Elementary differential equation theory will lead to a solution of (4.27), which is a wave equation. Green represents the solution by the equation:

$$w = f(ax+by+ct) = F(-ax+by+ct) \tag{4.30}$$

in the upper medium, with f representing the amplitude belonging to the incident wave and F that of the reflected plane wave. c is understood to be a negative quantity. In the lower medium Green writes:

$$w_1 = f_1(a_1 x + by + ct). \tag{4.31}$$

Substitution of these solutions into the equations yield solutions if $c^2 = \gamma^2(a^2+b^2)$ and $c^2 = \gamma_1^2(a_1^2+b^2)$. Application of (4.29) then gives:

$$\begin{aligned} f(by+ct)+F(by+ct) &= f_1(by+ct), \\ af'(by+ct)-aF'(by+ct) &= a_1 f'(by+ct) \end{aligned} \tag{4.32}$$

from which, taking the differential coefficient of the first equation and writing the characteristic only gives:

$$f' = F = f_1'$$

which in conjunction with the second equation of (4.32) yields:

$$\frac{F'}{f'} = \frac{1-a_1/a}{1+a_1/a} = \frac{a-a_1}{a+a_1} = \frac{\cot\theta - \cot\theta_1}{\cot\theta + \cot\theta_1} = \frac{\sin(\theta_1-\theta)}{\sin(\theta_1+\theta)} \quad (4.33)$$

which is Fresnel's sine law for light polarized in the plane of incidence, θ and θ_1 being, respectively, the angles of incidence and refraction.

Green then goes on from here, showing that if the generality of the wave function can be restricted to a function similar to that which describes the motion of a cycloid pendulum, certain interesting results connected with phase shifts in reflection can be demonstrated. Green's explanation of these restrictions and the derivation of the phase shifts are clear, and the reader is referred to the Green paper included in the readings, pp. 176–7.

The above argument leading to Fresnel's sine law should be sufficient to show how Green's theory is applied. It should also be taken as roughly equivalent to the mode of analysis which many dynamical aether theorists pursued during the nineteenth century, for it shows fairly clearly how the equations of motion and the boundary conditions are obtained and how additional hypotheses are incorporated in order to obtain optical results. The arguments are sophisticated, highly mathematical, and quite definite, and are not very different in spirit or level of physical competence from those presented in theories which have survived in physics until today.

I now turn to consider, rather briefly, Green's second case of polarization in which the plane wave is polarized in a direction perpendicular to the one just considered. Though this is the more interesting case—for it does not quite give Fresnel's tangent law and it also explicitly involves the problem of the longitudinal wave—it is considerably more complicated mathematically, and cannot be dealt with in detail without going beyond the scope and space limitations of this Commentary. The case is, of course, developed in the appended selections in Green's own words, and it has been critically considered by Lord Rayleigh (1871), to whose paper the reader may repair for an alternative account.

It is important, though, to comment on some of the physical assumptions made by Green in his second case, so that Green's position in the history of aether theories can be adequately considered.

When the light is polarized at right angles to the plane of incidence, $w = w_1 = 0$, and the boundary conditions become:

$$u = u_1 \quad v = v_1,$$

$$A\left(\frac{du}{dx} + \frac{dv}{dy}\right) - 2B\frac{dv}{dy} = A_1\left(\frac{du_1}{dx} + \frac{dv_1}{dy}\right) - 2B_1\frac{dv_1}{dy} \quad (4.34)$$

$$B\left(\frac{du}{dy} + \frac{dv}{dx}\right) = B_1\left(\frac{du_1}{dy} + \frac{dv_1}{dx}\right).$$

In accordance with what was said earlier, the B's may be cancelled in the last equation.

In this case, solution of the equations of motion yields two waves, a transverse wave with a velocity equal to $\sqrt{(B/\varrho)}$, and a longitudinal wave with a velocity of $\sqrt{A/\varrho}$. The two waves may be produced from a purely transverse wave by reflection of that wave at the interface $x = 0$.

The supposition the A/B equals a very great quantity is invoked, for reasons mentioned in the earlier discussion, to all but eliminate the effect of the longitudinal wave. This supposition, together with restrictions on the form of the wave function, identical to that referred to in connection with phase shifts in the first case, yields an expression for the relative intensities of the reflected to the incident wave of:

$$\frac{[(\mu^2+1)(\mu^2-a_1/a)+(\mu^2-1)^4(b^2/a^2)]^{\frac{1}{2}}}{[(\mu^2+1)^2(\mu^2+a_1/a)^2+(\mu^2-1)^4 b^2/a^2]^{\frac{1}{2}}} \quad (4.35)$$

where μ is the index of refraction and a and b are the amplitude coefficients of the x and y components of the wave function. Green shows that (4.35), as a first approximation, gives Fresnel's tangent formula, $\tan(\theta-\theta_1)/\tan(\theta+\theta_1)$. But Green's expression diverges sufficiently from the experimentally supported Fresnel expression for large θ's to render Green's theory inadequate in explaining this result.

Green's theory may be modified in various ways so as to eliminate this difficulty, however, the most successful of which was Lord Kelvin's modification. This was based on a reconsideration of Green's argument concerning the necessity to suppose the ratio of A to B infinitely great, and Kelvin was able to show that Green's approach, were A/B set equal to 0, could lead exactly to both of Fresnel's laws of reflection. I shall have more to say about Kelvin's views of Green's aether in a later section.

Green's aether theory was also suspect in that it would not, as presented above, account for double refraction, and Green developed a somewhat different second aether theory to explain this phenomenon. In this second theory he permitted the B terms to be functions of the strain direction in the doubly refracting medium. This was directly contrary to the assumption of the uniformity of B made in the paper we have considered, and Green never was able to effect an accommodation of his two aether theories.

Green's first theory was considered very successful, however, if a slight modification were made, in explaining the results of some subsequent experiments on reflection involving phase shifts and elliptical polarization which were done by Jamin (1850). Later Lord Rayleigh (1871), in assessing the merits of the modified Green's theory as against its various competitors, found that Green came off by far the best.

4. MacCullagh's Aether Theory

Green's aether theory had two major competitors in the elastic solid class during the nineteenth century. One was Cauchy's (1839) "third theory" which he developed in apparent ignorance of Green's earlier results. Cauchy commented on Green's theory in 1849, however, and disagreed with its approach, particularly on the usefulness and appropriateness of using the d'Alembert–Lagrange principle in investigating the optical aether. Cauchy's theory was subsequently developed by Haughton, St. Venant, and Sarrau.[†]

Green's other major competitor was a theory which had been developed in the years 1834–7 by James MacCullagh of Trinity College, Dublin.[‡] MacCullagh succeeded in putting his theory, which had been more a collection of hypotheses about the aether than a unified dynamical theory, on a relatively secure dynamical basis in 1839.

MacCullagh's theory has been highly praised by E. T. Whittaker in his *History of the Theories of the Aether and Electricity*. Whittaker wrote:

> MacCullagh . . . succeeded [in 1839] in placing his own theory, which all along had been free from reproach as far as agreement with optical experiments was concerned, on a sound dynamical basis; thereby effecting that reconciliation of the theories of light and dynamics which had been the dream of every physicist since the days of Descartes.
>
> The central feature of MacCullagh's investigation . . . is the introduction of a new type of elastic solid. He had in fact concluded from Green's results that it was impossible to explain optical phenomena satisfactorily by comparing the aether to an elastic solid of the ordinary type, which resists compression and distortion; and he saw the only hope of the situation was to devise a medium which should be as strictly conformable to dynamical laws as Green's elastic solid, and yet should have its properties specially designed to fulfil the requirements of the theory of light.

[†] See Glazebrook (1885), pp. 170–5, for a discussion and references.
[‡] A very similar theory was independently developed about this time by F. Neumann (1837). MacCullagh's theory is sometimes referred to as the MacCullagh–Neumann theory.

I will discuss MacCullagh's new type of elastic solid shortly. It might be well to note here though the reason which Whittaker offers for MacCullagh's theory being largely ignored in the nineteenth century. Whittaker wrote:

> MacCullagh's work was regarded with doubt by his own and the succeeding generation of mathematical physicists, and can scarcely be said to have been properly appreciated until Fitzgerald drew attention to it forty years afterwards. But there can be no doubt that MacCullagh really solved the problem of devising a medium whose vibrations, calculated in accordance with the correct laws of dynamics, should have the same properties as the vibrations of light.
>
> The hesitation which was felt in accepting the rotationally elastic aether [i.e. MacCullagh's medium] arose mainly from the want of any readily conceived example of a body endowed with such a property [i.e. purely rotational elasticity]. This difficulty was removed in 1889 by Sir William Thompson (Lord Kelvin) who devised mechanical models possessed of rotational elasticity.

I have quoted this much of Whittaker on MacCullagh because of Whittaker's influence on the views which many contemporaries have of nineteenth-century aether theories. I shall show below that Whittaker's characterization of MacCullagh's aether as being in accord with the laws of dynamics is incorrect, and the reasons why MacCullagh's aether was not acceptable to nineteenth-century physicists are not those which Whittaker cites. In order to understand the actual historical relations between the various aether theories in the nineteenth century, as well as to become somewhat clearer on just how successful mechanical explanations can be in explicating optical theories, I will consider MacCullagh's aether in some detail.

MacCullagh relates in his 1839 paper that certain laws of the reflection and refraction of light at the surface of crystals, about which he had previously written, and which were, he claimed, "remarkable for their simplicity and elegance, as well as for their agreement with exact experiments", were, none the less, wanting a coherent mechanical explanatory basis. In the 1839 paper, of which I have included some extracts, MacCullagh gave a theory

THE ELASTIC SOLID AETHER 61

which he believed to be adequate for explaining reflection, refraction, and double refraction. The theory is based on two assumptions. MacCullagh hypothesized:

> *First*, that the density of the luminiferous aether is a constant quantity; in which it is implied that this density is unchanged either by the motions which produce light or by the presence of material particles, so that it is the same within all bodies as in free space, and remains the same during the most intense vibrations. *Second*, that the vibrations in a plane wave are rectilinear, and that, while the plane of the wave moves parallel to itself, the vibrations continue parallel to a fixed right line, the direction of this right line and the direction of a normal to the wave being functions of each other. This supposition holds in all known crystals, except quartz, in which the vibrations are elliptical.

The first assumption will, in effect, both rule out any compressional wave, as well as require that refraction be made dependent, as it was for Huygens, but not for Young, Fresnel, or Green, on the difference in the rigidities of two media. The second assumption will be utilized in the derivation of the potential energy function of the aether.

MacCullagh's dynamical approach is roughly the same as Green's, and though Whittaker suggests that MacCullagh was aware of Green's work, I have not found any evidence that this is so. Stokes (1862), in commenting on MacCullagh's mode of analysis, also believes that MacCullagh was unaware of Green's aether theory at this time.

Like Green, MacCullagh applies the d'Alembert–Lagrange principle to his aether and seeks to determine the appropriate energy function which would satisfy the various restrictions he has imposed on his medium. MacCullagh uses the symbol V in place of Green's ϕ, and, of course, its determinate form will also differ because of their different views of the aether. As MacCullagh notes, the form of V is dependent "on the assumptions stated respecting the ethereal vibrations...".

The general form of V is, of course, given as usual by the general variational equation (4.14), which in MacCullagh's symbolism becomes:

$$\iiint dx\, dy\, dz \left(\frac{d^2\xi}{dt^2}\delta\xi + \frac{d^2\eta}{dt^2}\delta\eta + \frac{d^2\zeta}{dt^2}\delta\zeta \right) = \iiint dx\, dy\, dz\, \delta V \quad (4.36)$$

with x, y, and z being the coordinates of an aether particle before it is disturbed, and $x+\xi$, $y+\eta$, and $z+\zeta$ its coordinates at time t. MacCullagh also sets the aether density, which is in his theory everywhere the same, equal to unity so that $dxdydz$ may represent either an element of volume or of mass.

The determinate form of V is obtained by considering a system of plane waves moving through the aether, parallel to which we construct a plane $x'y'$. The waves are apparently polarized parallel to the y' axis, so that the disturbance of an aether particle is confined to the y' direction, and ξ' and ζ' are both equal to zero. Then an elementary differential parallelpiped is constructed, with sides $dx'dy'dz'$ respectively parallel to the axes of $x'y'z'$. I have drawn the parallelpiped in Fig. IV.1, and attempted to represent the effect of the plane wave moving up the z axis on several "slices" of aether particles. As a result of the passage of the wave, the bottom of the differential volume will be shifted parallel to the $x'y'$ plane with respect to the top of the volume. Consequently, a line connecting the top corner with the bottom corner, formerly directly beneath it, will no longer be parallel to the z' axis, but will be inclined to it at an angle k where $\tan k = d\eta'/dz'$.

MacCullagh then argues that the function V for which he is seeking: "can only depend upon the directions of the axes of $x'y'z'$ with respect to fixed lines in the crystal, and upon the angle k which measures the change of form produced in the parallelpiped by vibration."

This is the most general supposition which can be made concerning it. MacCullagh then uses his second assumption, quoted on p. 61 above, which implies that any one direction, say x', determines the other two directions y' and z' because of the imposed requirement of constant orthogonality. Thus V can be written as a function of k and x' alone.

In the mathematical section II of his paper (which is not included in these selections) MacCullagh shows that analytical geometrical

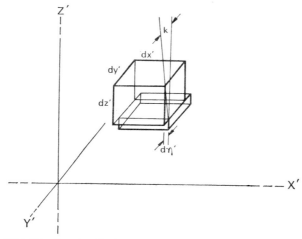

FIG. IV.1. The shearing displacement of a segment of a differential volume of MacCullagh's aether, caused by a plane polarized wave moving in the positive Z axis direction.

considerations associated with arbitrary rotations of coordinate systems allow him to demonstrate that the angle k and the direction of x' with respect to the primary xyz coordinate system are known if some special quantities XYZ are known. These quantities XYZ are curl functions of the displacement ξ, η, ζ of the aether particle considered above on p. 62, and are defined in Cartesian component terms as:

$$X = \frac{d\eta}{dz} - \frac{d\zeta}{dy} \quad Y = \frac{d\zeta}{dx} - \frac{d\xi}{dz} \quad Z = \frac{d\xi}{dy} - \frac{d\eta}{dx} \quad (4.37)$$

V therefore, according to MacCullagh, may be considered a function of XYZ alone.

MacCullagh's argument is not completely cogent, however, and in 1862 in an influential report to the British Association for the Advancement of Science, G. G. Stokes criticized MacCullagh's derivation of V, noting:

> [MacCullagh's] reasoning, which is somewhat obscure, seems to me to involve a fallacy. If the form of V were known, the rectilinearity of vibration and the constancy in the direction of vibration for a system of plane waves travelling in any given direction would follow as a *result* of the solution of the problem. But in using equation . . . [(4.36)] we are not at liberty to substitute for V an expression which represents that function *only on the condition that the motion be what it actually is*, for we have occasion to take the variation δV of V, and this variation must be the most general that is geometrically possible though it be dynamically impossible. That the form of V arrived at by MacCullagh, is inadmissible, is I conceive, proved by its incompatibility with the form deduced by Green from the very same supposition of the *perfect* transversality of the transversal vibrations; for Green's reasoning is perfectly straightforward and irreproachable. Besides MacCullagh's form leads to consequences absolutely at variance with dynamical principles.

I shall comment on the dynamical deficiencies of MacCullagh's aether shortly; for now it will suffice to sketch the remaining steps of MacCullagh's argument.

This can be done quickly as it is very similar to Green's more general case. Supposing k very small, XYZ will be very small, and V can be expanded in a series. First-order terms ought to vanish assuming an initially unstrained medium, and third- and higher-order terms are neglected in comparison with the second order. Since we now have a second-degree function of three quantities—recall that Green had six—V_2 will be a homogeneous function containing in its most general form terms involving the squares and products of XY and Z with six arbitrary coefficients. The coefficients associated with the product terms can be made equal to zero by choosing the proper orientation of the XYZ axes, since the quantities XYZ transform in the same manner as do axes, whence MacCullagh obtains:

$$V = -\tfrac{1}{2}(a^2X^2 + b^2Y^2 + c^2Z^2) \tag{4.38}$$

in which $-\frac{1}{2}a^2$, $-\frac{1}{2}b^2$, and $-\frac{1}{2}c^2$ are the arbitrary coefficients. The negative sign is introduced so that the velocity of propagation can never become imaginary.

Having arrived at the form of V MacCullagh says: "...we may now take it for the starting point of our theory, and dismiss the assumptions by which we were conducted to it." He then develops, in a manner not unlike Green's, the equations of motion, the boundary conditions, and based on these, derivations of Fresnel's sine and tangent laws, which he obtains exactly.

Interesting though MacCullagh's theory may be, it is not a dynamical theory in the same sense that Green's is. It was shown by Stokes (1862) and also Lorentz (1901) that the MacCullagh aether violates the dynamical principle of the equality of action and reaction in regard to moments. In Stokes' (1862) words:

> The condition of moments is violated. It is not that the resultant of the forces acting on an element of the medium does not produce its proper momentum in changing the motion of translation of the element . . . but that a couple is supposed to act on each element to which there is no corresponding reacting couple.

It might also be noted here that MacCullagh's theory was also refuted by experiment. Lorenz (1861) and later Lord Rayleigh (1871) showed that an aether which assumed constant density, as did MacCullagh's and Neumann's, implied the existence of two polarizing angles at $\pi/8$ and $3\pi/8$ radians, whenever the difference in the indices of refraction between two media is small. Experiments disclose only one such angle, however, and imply that MacCullagh's theory is incorrect.

Stokes' objection against the theory was generally accepted, and eliminated MacCullagh's theory from serious consideration as a dynamical theory. I shall have occasion to quote Larmor on this point in the next chapter.

MacCullagh, contrary to Whittaker's implications, never thought he had provided a satisfactory dynamical theory of the aether. At the conclusion of his 1839 paper MacCullagh wrote:

> In this theory, everything depends on the form of the function V; and we have seen that, when that form is properly assigned, the laws by which crystals act upon light are included in the general equation of dynamics. This fact is fully proved by the preceding investigations. But the reasoning which has been used to account for the form of the function is indirect, and cannot be regarded as sufficient, in a mechanical point of view. It is, however, the only kind of reasoning which we are able to employ, as the constitution of the luminiferous medium is entirely unknown.

In a sense, the MacCullagh aether can be defended if it can be supplemented with *another* aether which would provide the restoring couple missing from the MacCullagh aether. This supplementation in fact is the case in models which Kelvin constructed for Green's aether in 1889 and 1890 and which Whittaker erroneously implies were realizations of MacCullagh's aether. This is a subtle point though, and I shall return to it below when I consider Kelvin's models in detail.

The MacCullagh aether is more important as an electromagnetic aether than it is as a dynamical aether. As we shall see in the next chapter, both Fitzgerald and Larmor explicitly used generalized models of the MacCullagh aether in terms of which to interpret Maxwell's electromagnetic theory.

The idea of the aether as an elastic solid was seriously pursued and developed throughout the nineteenth century until about 1890. The contributions of the many physicists who formulated various aether theories are too many and various to do little more than mention.

Stokes' more positive contributions to aether theory ought to be cited. In addition to working out the theory of aberration which was discussed in the previous chapter, Stokes also formulated a dynamical theory of diffraction (1849) which was based on Green's theory of the aether. Stokes also developed a theory of the "fluid" aether in which he introduced an important distinction between the "rigidity" and the "plasticity" of a substance. With this distinction, Stokes was able to explain how the planets could move easily through the aether, the aether behaving in this case as a fluid, while at the same time light might be rapidly propagated through it, as if it were a rigid body. Glazebrook (1885) commented on this theory of Stokes, noting that: "These same views also tend to confirm the belief that for fluids, and among them the aether, the ratio of A to B (the elastic constants of the medium in Green's notation) will be extremely great". The aether, according to Stokes, acts very much like a synthetic plastic, which is often sold for amusement purposes, and sometimes known as "Silly Putty", or "Monster Putty". This substance shatters like glass when struck sharply, but flows like a liquid when subjected to a constant force over a long period of time.

I have previously mentioned some of Lord Rayleigh's contributions to aether theory; he made a number of others. On the Continent, G. Kirchhoff (1876) pursued an interesting aether theory and also developed a particularly elegant analysis of diffraction which, however, we cannot discuss for lack of space.

Several chroniclers of nineteenth-century aether theories[†] find it useful to make a distinction between theories like Green's, MacCullagh's, and Cauchy's, which are elastic solid theories, and a different type of aether theory which developed primarily during the 1870s and 1880s. This new type of aether theory took careful cognizance of the interaction between aether and matter in ways which the earlier theories had overlooked. Communication of momentum between the aether and the matter it interpenetrated had not been dealt with by these earlier theories in any thorough manner.

Boussinesq (1867) was perhaps the first aether theorist to consider this problem seriously and attempt to explain reflection, refraction, polarization, dispersion, etc., on the basis of a uniform aether which varied in rigidity and density only when it got entangled

[†] These are primarily Glazebrook (1885) and Basset (1892). Whittaker is not among this group, and his *History* is rather deficient in this regard, though he does discuss Boussinesq's (1867) aether theory. See below for more comments on this point.

with matter. (Actually the properties of the aether *itself* do not change, but the net effect is as they did.)

Other theories of this type were vigorously developed by Sellmeier (1872), Helmholtz (1875), Lommel (1878), Voigt (1883), and Ketteler (1885).[†] The most important example which belongs to this type, even though it was electromagnetic rather than optical in nature, is the Lorentz aether which will be discussed in Chapter VI.

5. Kelvin's Aether and his Models

In England one of the proponents of this new type of aether theory was Lord Kelvin. In 1884 Kelvin gave a series of high-level lectures on aether theories at the Johns Hopkins University in Baltimore. The lectures were published shortly thereafter as *Baltimore Lectures on Molecular Dynamics and the Wave Theory of Light*. These lectures constituted an inquiry into the dynamical shortcomings of various aethers, and focused on Green's theory. Kelvin told his "coefficients", as he punningly termed the twenty-one professors who attended his lectures, that Green's theory at least had to be supplemented to include the interaction with atoms in ponderable matter. In his own characteristic style, Kelvin presented various mechanical models consisting of shells and springs whose action might mimic in a "rude" manner the hypothetical interaction of aether and matter.

Kelvin was by no means satisfied with the Green aether, even as supplemented with such notions, but he trusted it far more than he did the electromagnetic theory of light which was at the time receiving more and more attention, even though Hertz' important experiments had as yet not been performed. Though there were some nineteenth-century aether theories which were strongly influenced by Maxwell's electromagnetic theory and which will be dis-

[†] See Glazebrook (1885) for a discussion of these theories and for references.

cussed in the next two chapters, many late nineteenth-century aether theories were developed entirely independently of any relation to electromagnetism, or to Maxwell's theory, specifically. Kelvin's antipathy to Maxwell's theory is well known. In the Baltimore lectures he wrote:

> If I knew what the electromagnetic theory of light is, I might be able to think of it in relation to the fundamental principles of the wave theory of light. But it seems to me that it is a rather backward step from an absolutely definite mechanical notion that is put before us by Fresnel and his followers to take up the so-called Electro-magnetic theory of light in the way it has been taken up by several writers of late. . . . I merely say this in passing, as perhaps some apology is necessary for my insisting upon the plain matter-of-fact dynamics and the true elastic solid as giving what seems to me the only tenable foundation for the wave theory of light in the present state of our knowledge.

The 1884 version of the *Baltimore Lectures* was somewhat inconclusive on the positive side, as it raised more problems that it solved. As Glazebrook (1885) noted, however, it did develop an interest in England in the interaction type of aether theories.

Four years later Kelvin (1888) came upon a most important theoretical discovery regarding aether theory. It is worthwhile to let him tell it in his own words:

> Since the first publication of Cauchy's work on the subject in 1830, and of Green's in 1837, many attempts have been made by many workers to find a dynamical foundation for Fresnel's laws of reflexion and refraction of light, but all hitherto ineffectually. On resuming my own efforts since the recent meeting of the British Association in Bath, I first ascertained that an inviscid fluid permeating among pores of an incompressible, but otherwise sponge-like, solid, does not diminish, but on the contrary augments, the deviation from Fresnel's law of reflexion for vibrations in the plane of incidence. Having thus, after a great variety of previous efforts which had been commenced in connexion with preparations for my Baltimore Lectures of this time four years ago, seemingly exhausted possibilities in respect to *incompressible* elastic solid, without losing faith either in light or in dynamics, and knowing that the condensational-rarefactional wave disqualifies any elastic solid of *positive* compressibility, I saw that nothing was left but a solid of such negative compressibility as should make the velocity of the condensational-rarefactional wave zero. So I tried

> it and immediately found that it, with other suppositions unaltered from Green's, exactly fulfils Fresnel's "tangent-law" for vibrations *in* the plane of incidence, and his "sine-law" for vibrations perpendicular to the plane of incidence. I then noticed that homogeneous air-less foam held from collapse by adhesion to a containing vessel, which may be infinitely distant all round, exactly fulfils the condition of zero velocity for the condensational-rarefactional wave; while it has a definite rigidity and elasticity of form, and a definite velocity of distortional wave, which can easily be calculated with a fair approximation to absolute accuracy.
>
> Green, in his original paper "On the Reflexion and Refraction of Light", had pointed out that the condensational-rarefactional wave might be got quit of in two ways, (1) by its velocity being infinitely small, (2) by its velocity being infinitely great. But he curtly dismissed the former and adopted the latter, in the following statement:—"And it is not difficult to prove that the equilibrium of our medium would be unstable unless $A/B > 4/3$. We are therefore compelled to adopt the latter value of A/B" (∞) "and thus to admit that in the luminiferous ether, the velocity of transmission of waves propagated by normal vibrations, is very great compared with that of ordinary light." Thus originated the "jelly" theory of ether, which has held the field for fifty years against all dynamical assailants, and yet has failed to make good its own foundation.
>
> But let us scrutinize Green's remark about instability. Every possible infinitesimal motion of the medium is, in the elementary dynamics of the subject, proved to be resolvable into coexistent condensational-rarefactional wave-motions. Surely, then, if there is a real finite propagational velocity for each of the two kinds of wave-motion, the equilibrium *must* be stable! And so I find Green's own formula proves it to be *provided we either suppose the medium to extend all through boundless space, or give it a fixed containing vessel as its boundary.*

Kelvin's analysis thus sets Green's coefficient A equal to zero, and proves that the medium is not unstable to the point of explosion, as was thought by Green, if Kelvin's conditions of a containing vessel or infinite extent of the aether hold. This type of aether will not support a longitudinal wave, for as regards its energy distribution:

> If $A = 0$, as we are going to suppose for our optical problem, no work is required to give the medium any infinitely small irrotational displacement; and thus we see the explanation of the zero velocity of the condensation and rarefactional wave . . . (Kelvin, 1888).

Such an aether is sometimes referred to as a quasi-labile aether, inasmuch as it is "labile" with respect to compression, much as a cylinder rolling on a horizontal plane is in labile equilibrium. The aether is, however, resistant to *rotational* forces.†

In 1889 and 1890 Kelvin was able to develop gyrostatic mechanical models of such a quasi-labile aether. In Kelvin's (1890) words, such a model was a "mechanical realization of the medium to which I was led one and one half years ago from Green's original theory by purely optical reasons".

I have included Kelvin's article which discussed this "mechanical realization" in order to give the flavor of Kelvin's "model" type of thinking, and in order to contrast it with his more theoretical investigations which are not very different in spirit from those of the typical nineteenth-century aether theorist.

Unfortunately, Kelvin is often taken, when he is discussing a "mechanical realization" of his theories, as being typical of nineteenth-century British thinking on the aether, and this has led to a number of confusions, especially in contemporary philosophy of science, about the relation between model and theory and the reality status of theories.

Kelvin's model is important to the extent that it made a mechanical explanation of the optical aether more plausible by showing that there was nothing inconsistent in those mechanical theories which were characterized by a gyrostatic rigidity. Green's theory as modified by Kelvin has this peculiar property, as does MacCullagh's aether theory about which I spoke earlier. We shall see that the Kelvin model also, however, assumes a second interpenetrating aether against which the rotational torques of the optical aether are exerted, thus outflanking Stokes' objection against the MacCullagh aether. Larmor (1894) showed that the Green–Kelvin aether and the MacCullagh aether were intertranslatable, thus proving formally what the model might lead one to suspect.

† Soon after Kelvin's theory appeared, R. T. Glazebrook (1888) applied it to the phenomenon of double refraction and obtained satisfactory results, thus bringing together, for the first time, Green's two aethers. (See above, p. 58.)

Kelvin's gyrostatic aether "realization" is extremely complicated, and unfortunately Kelvin's brusque manner of presenting it does not do much to clarify the matter. I intend only to sketch the principle features of his structure, and must refer the reader who desires a thorough analysis of the model from the point of view of theoretical gyroscopic mechanics to Lorentz' lecture (1901) on Kelvin's model.

Kelvin's model consists of a three-dimensional array of connected tetrahedrons, essentially built up on the basis of the plane system of equilateral triangles shown in Fig. IV.2. If the nonshaded triangles in the figure are taken as the bases of the tetrahedrons, we can consider *PQRS* and *T* as the top vertices of these

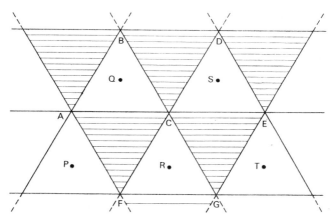

FIG. IV.2. A schematic representation of Kelvin's gyrostatic aether. (After Lorentz, 1901.)

tetrahedrons, themselves constituting the base points of a second level of tetrahedrons. If we carry out the building up and out of these levels systematically, every corner point in the system will be the common vertex of four tetrahedrons.

To imagine the mode of connection of these tetrahedrons which will give him the labile property of non-resistance to compression,

Kelvin proposes that at each vertex there is a "ball and twelve socket mechanism" from which issue "six fine straight rods and six straight tubes, all of the same length, the internal diameter of the tubes exactly equal to the external diameter of the rods". The bars issuing from one of the balls fit into and slide without friction in the tubes of the other balls, and vice versa, the interconnected tubes and rods thus now constituting the edges of the tetrahedrons. This interconnected system is the *framework* into each tetrahedron of which Kelvin then introduces "a rigid frame G [consisting] of three rods [which can expand or contract] fixed together at right angles to one another through one point O". This G frame is so positioned that three of its bars are put into permanent but sliding frictionless contact with the three pairs of rigid sides of any tetrahedron of the framework. Lorentz (1901) notes that "in a regular tetrahedron these bars coincide with the lines joining the midpoints of the opposite edges, but also in any tetrahedron whatever, a set of mutually orthogonal intersecting lines joining pairs of opposite edges can always be assigned". Kelvin then proposes to proliferate G frames throughout his framework so that the G frames constitute a "second homogeneous assemblage".

It can be shown that if the system of framework and G frames is subjected to an infinitely small homogeneous *irrotational* distortion, that the G frames do not undergo any rotation, though they do translate. Kelvin argues, however, that should the distortion have a *rotational* component, such as might be produced by an arbitrary displacement of the system, that: "any infinitely small homogeneous displacement whatever of the primary assemblage [i.e. the framework] produces a rotation of each frame equal to and round the same axis as its own rotational component." This is a most important property since if a resistance to rotation alone can be conferred on the G frames, the property of being both labile for compression and at the same time resistant to rotation will exist in this system. This is exactly what Kelvin sets out to do.

He introduces resistance to rotation into the G frames by mounting two gyrostats on each bar of the G frame. One such gyrostat of the solid type is depicted in Fig. IV.3. Kelvin was not particular about which type of gyrostat to use, as this was only a model, and he also describes a liquid gyrostat in section 12 of the appended selection.

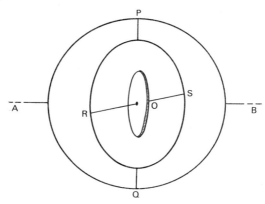

FIG. IV.3. A solid gyrostat. Line AB is in the axis of the G frame. The solid rotating flywheel at O is free to turn on axis RS, while the inner ring may rotate about axis $PQ \perp AB$. The outermost ring is fixed in the G frame. (After Lorentz, 1901.)

What the gyrostats do, when six of the solid or twelve of the liquid variety are introduced into each G frame, is to provide the necessary resistance to rotation but not to translation of the G frame. Consequently the system as a whole exhibits the types of properties ascribed to both the Green and MacCullagh aethers. Because of the existence of *framework*, however, the G frames can react on another object, and thus outflank Stokes' criticism against the MacCullagh aether.

It does not appear that Kelvin thought of this model as anything more than an analogy. He did maintain a faith in a modified Green aether until at least 1904, though from 1899 to about 1902 he thought he might have to relinquish this aether since it led to some unfortunate predictions concerning the amount of energy dissipated in the case of a vibrating sphere in the aether. In 1902, however, Kelvin discovered another modification of the Green aether which permitted him to revise these predictions. The story is told in the 1904 edition of the *Baltimore Lectures* and cannot be discussed here.

Kelvin's approach to aether theory raises, as I suggested above, some important philosophical questions. One which has not been adequately considered is the question of effecting a reduction of one theory or one science to another.† Most nineteenth-century aether theorists were seeking a reduction of the phenomena and laws—and occasionally even theory—of optics to some form of mechanical theory. Such a mechanical theory need not be much more "concrete" than the Green theory or the Kelvin–Green theory. Effecting a reduction to mechanics or—what is the same thing—finding a general explanation in terms of a mechanical theory, must be distinguished from imagining some concrete model of wheels, pulleys, gyroscopes, or whatever. Providing the latter provides an *analogy* and not a *reduction*. The nineteenth-century aether theorists understood the distinction quite clearly, and when they began to consider a mechanical explanation of Maxwell's theory, became quite self-conscious about the criteria of "mechanical explanations". We turn to consider such theories next.

---

† See Nagel (1961) and Schaffner (1967) for analyses of reduction.

CHAPTER V

THE ELECTROMAGNETIC AETHER

IN THE previous chapter I mentioned the lack of contact between some late nineteenth-century optical aether theories and Maxwell's electromagnetic theory. There are some reasons for this. Certainly prior to Hertz' (1888) important experimental production and detection of Maxwell's electromagnetic waves, Maxwell's theory was only one of many competing *optical* theories, and occupied a similar non-paramount position with respect to electromagnetic theories. We have discussed the various optical theories in the last chapter. With respect to competing electromagnetic theories, Maxwell's theory was flanked on the right by Kelvin's various attempts to characterize the electric and magnetic aethers, on the left by various continental action at a distance schools, the most prominent of which was Weber's, and from above (to extend the metaphor) by Helmholtz' influential synthetic theory. This book cannot by its very nature consider very carefully these complex competing doctrines of electromagnetic action, and must refer the reader to other literature.† The purpose of this chapter is to sketch some of the notions of the aether that were held by Faraday, Maxwell, and some of Maxwell's followers, with emphasis on the latter, and to show how theories of the mechanical optical aether were eventually brought into close relations with Maxwell's theory.

---

† See Whittaker (1960), Chaps. 7–10, Thomson (1885), and Rosenfeld (1957).

1. Faraday

Michael Faraday was persuaded by his own doctrine of "lines of force" that the mode of connection between ponderable bodies that were coupled by electrical, magnetic, gravitational, or optical interaction, was via a peculiar contact action of the bodies themselves. Bodies, for Faraday, were aggregates of Boscovichian atoms which extended indefinitely outward into space. Lines of force were apparently related to this extension, according to the statements of Faraday which I shall cite below.

Faraday was quite clear about his antipathetic view of the Fresnel and post-Fresnel optical aethers. In his seminal essay, "Thoughts on Ray Vibrations", published in 1846, Faraday wrote:

> The point intended to be set forth . . . was, whether it was not possible that the vibrations which in a certain theory are assumed to account for radiation and radiant phenomena may not occur in the lines of force which connect particles, and consequently masses of matter together; a notion which as far as it is admitted, will dispense with the aether, which, in another view, is supposed to be the medium in which the vibrations take place.

Faraday, accordingly, was against the aether as well as being opposed to the "action at a distance" approach. He is, however, committed to a "medium", if this word can be used for the bodies themselves, through which electric and magnetic action travels, and travels with a velocity comparable with the velocity of light. Faraday thus suggested the identity of light waves and electromagnetic waves. Faraday has also been looked on as being the originator of "field theory", in so far as Maxwell's electromagnetic field theory follows closely both on Faraday's experimental work and the mathematization of some of Faraday's speculative ideas concerning the lines of force and the electrotonic state. This latter notion was a theoretical idea in terms of which Faraday thought he might account for electromagnetic induction.[†]

[†] See L. P. Williams' (1965) excellent study on Faraday, and also Tricker (1966) for additional material on the electrotonic state.

Faraday contended that he was led to this possibility of an electromagnetic view of light from the Boscovichian idea of matter. This notion, in Faraday's mind, considered atoms not as "so many little bodies surrounded by forces, ... these little particles [having] a definite form and a certain limited size"—rather, in the Boscovichian atom "that which represents size may be considered as extending to any distance to which the lines of force of the particle extend: the particle indeed is supposed to exist only by these forces, and where they are it is".

2. Maxwell

Faraday's view of the interaction of charged and magnetized bodies via lines of force and the possible identification of light with electromagnetic vibrations strongly influenced James Clerk Maxwell. In a paper written in late 1855 Maxwell (1856) first presented some of his thoughts on Faraday's lines of force, offering certain mathematical expressions drawn from fluid mechanics partly based on some of Lord Kelvin's work, in terms of which to interpret the "lines of force". For example, in his section on Faraday's electrotonic state, Maxwell defined certain complicated functions which might characterize this state. The quantity which represents the electrotonic intensity turns out to be identical with the contemporary vector potential a, which is related to the magnetic induction vector by: curl $a = B$. Maxwell had little faith in this tentative mathematization of Faraday's theory, however, and noted in his concluding remarks that:

> In these . . . laws I have endeavored to express the idea which I believe to be the mathematical foundation of the modes of thought indicated in the *Experimental Researches*. I do not think that it contains even the shadow of a true physical theory; in fact its chief merit as a temporary instrument of research is that it does not, even in appearance, *account* for anything.

Maxwell's mode of analysis thus far had little contact with the aether theories which were discussed in the previous chapters.

Over the next ten years, however, Maxwell developed theories of a mechanical-electromagnetic aether and a purified electromagnetic aether, both of which he *identified* at different times with the optical aether.

The evolution of Maxwell's thought on the aether can only be sketched in broad outlines. The interested reader should refer to Whittaker (1960) and to a recent paper by Joan Bromberg (1968) for more detail.

Maxwell published his first electromagnetic theory of light in 1861–2 in a series of papers that appeared in the *Philosophical Magazine* under the title "On Physical Lines of Force". This first theory is, as Bromberg (1968) points out, not really an "electromagnetic theory of light". Rather it is "better characterized as an electro-mechanical theory of light, for in it the equations of light are derived, not from electromagnetic laws alone, but partly from electromagnetic laws and partly from laws of mechanics".

Maxwell's 1861–2 mechanical model of the electromagnetic aether is well known, if not very well understood, and is adequately treated in a volume in this series by Tricker (1966). Suffice it to say here that the electromagnetic theory of *light* does not appear until the model of whirling magnetic vortices and electrically charged idle wheels, which was adequate for characterizing the relations between currents and magnetism, is altered to incorporate a representation of a static electrical field. This was done by now conceiving of the whirling vortices as static cells which are twisted from their equilibrium position by the tangential displacement of the charged idle wheels. This displacement is occasioned by an electromotive force applied to the system, and is Maxwell's well-known "electric displacement", a change of which (because of the motion of "charges") constitutes the "displacement *current*". Elimination of the applied field permits the medium to return to its original undisplaced untwisted state.

Maxwell offered a kind of defense for the ascription of elasticity to these cells which foreshadows his later identification of the electromagnetic and optical media. Early in part III of his paper he wrote:

> The substance in the cells possesses elasticity of figure . . . similar to that observed in solid bodies. The undulatory theory of light requires us to admit this kind of elasticity in the luminiferous medium in order to account for transverse vibrations. We need not then be surprised if the magnetoelectric medium possesses the same property.

After incorporating this into his theory, Maxwell calculated "the rate of propagation of transverse vibrations through the elastic medium of which the cells are composed, on the supposition that its elasticity is due entirely to forces acting between pairs of particles". He found that it "agrees so exactly with the velocity of light calculated from ... optical experiments ...that we can scarcely avoid the inference that *light consists in the transverse undulations of the same medium which is the cause of electric and magnetic phenomena*". (Emphasis is Maxwell's.) Maxwell thus identified the two aethers.

Maxwell was apparently uneasy with the mechanico-electrical system on which his theory depended, and in late 1864 represented his theory on a rather different basis, emancipating it from the earlier model that aided its formulation. He did not, however, consider that he had in any way eliminated the aether, which was now conceived of as being accurately characterized by his electromagnetic field equations.

It is worth quoting Maxwell rather extensively on this point. At the beginning of his paper published in early 1865 Maxwell wrote:

> [Rather than seeking an action at a distance theory] I have preferred to seek an explanation of electrostatic and [electromagnetic phenomena] by supposing them to be produced by actions which go on in the surrounding medium as well as in the excited bodies, and endeavouring to explain the action between distant bodies without assuming the existence of forces capable of acting directly at sensible distance.
>
> (3) The theory I propose may therefore be called a theory of the *Electromagnetic Field*, because it has to do with the space in the neighbourhood of the electric or magnetic bodies, and it may be called a *Dynamical* Theory,

because it assumes that in that space there is matter in motion, by which the observed electromagnetic phenomena are produced.

(4) The electromagnetic field is that part of space which contains and surrounds bodies in electric or magnetic conditions.

It may be filled with any kind of matter, or we may endeavour to render it empty of all gross matter, as in the case of Geissler's tubes and other so-called vacua.

There is always, however, enough of matter left to receive and transmit the undulations of light and heat, and it is because the transmission of these radiations is not greatly altered when transparent bodies of measurable density are substituted for the socalled vacuum, that we are obliged to admit that the undulations are those of an aethereal substance, and not of the gross matter, the presence of which merely modifies in some way the motion of the aether.

We have therefore some reason to believe, from the phenomena of light and heat, that there is an aethereal medium filling space and permeating bodies, capable of being set in motion and of transmitting that motion from one part to another, and of communicating that motion to gross matter so as to heat it and affect it in various ways.

(5) Now the energy communicated to the body in heating it must have formerly existed in the moving medium, for the undulations had left the source of heat some time before they reached the body, and during that time the energy must have been half in the form of motion of the medium and half in the form of elastic resilience. From these considerations Professor W. Thomson has argued, that the medium must have a density capable of comparison with that of gross matter, and has even assigned an inferior limit to that density.

(6) We may therefore receive, as a datum derived from a branch of science independent of that with which we have to deal, the existence of a pervading medium, of small but real density, capable of being set in motion, and of transmitting motion from one part to another with great, but not infinite, velocity.

Hence the parts of this medium must be so connected that the motion of one part depends in some way on the motion of the rest; and at the same time these connexions must be capable of a certain kind of elastic yielding, since the communication of motion is not instantaneous, but occupies time.

The medium is therefore capable of receiving and storing up two kinds of energy, namely, the "actual" energy depending on the motions of its parts, and "potential" energy, consisting of the work which the medium will do in recovering from displacement in virtue of its elasticity.

The propagation of undulations consists in the continual transformation of one of these forms of energy into the other alternately, and at any instant the amount of energy in the whole medium is equally divided, so that half is energy of motion, and half is elastic resilience.

(7) A medium having such a constitution may be capable of other kinds of motion and displacement than those which produce the phenomena of light and heat, and some of these may be of such a kind that they may be evidenced to our senses by the phenomena they produce.

The theory which Maxwell presented in a systematic form in 1865 contained twenty equations in twenty unknowns. Even this form, however, which made use of the vector and scalar potentials in its basic equations, could have been somewhat "simplified" had Maxwell used vector equations rather than writing out the equations component-wise. Nevertheless, the theory was exceedingly complicated, and Maxwell's mode of presentation did not add to the clarity. There is sufficient testimony on this matter by prominent scientists, such as Ehrenfest (1916) and Sommerfeld (1933), who have commented on the difficulties of reading Maxwell in the original, and it seems that the simplification of his theory by Heaviside and Hertz in 1889 and 1890 had almost as much impact on the acceptance of the Maxwell theory as did Hertz' experiments.

The theory which Maxwell gave in 1865 is non-mechanical, though it is clearly not anti-mechanical. Maxwell's difference in orientation in his two papers (i.e. 1862 and 1865) is brought out very clearly by comments which he makes near the end of Part III of his later paper. Maxwell (1865) wrote:

(73) I have on a former occasion attempted to describe a particular kind of motion and a particular kind of strain, so arranged as to account for the phenomena. In the present paper I avoid any hypothesis of this kind; and in using such words as electric momentum and electric elasticity in reference to the known phenomena of the induction of currents and the polarization of dielectrics, I wish merely to direct the mind of the reader to mechanical phenomena which will assist him in understanding the electrical ones. All such phrases in the present paper are to be considered as illustrative, not as explanatory.

(74) In speaking of the Energy of the field, however, I wish to be understood literally. All energy is the same as mechanical energy, whether it exists in the form of motion or in that of elasticity, or in any other form. The energy in electromagnetic phenomena is mechanical energy. The only question is, Where does it reside? On the old theories it resides in the electrified bodies, conducting circuits, and magnets, in the form of an un-

known quality called potential energy, or the power of producing certain effects at a distance. On our theory it resides in the electromagnetic field, in the space surrounding the electrified and magnetic bodies, as well as in those bodies themselves, and is in two different forms, which may be described without hypothesis as magnetic polarization and electric polarization, or, according to a very probably hypothesis, as the motion and the strain of one and the same medium.

Maxwell's theory was clearly a kind of aether theory, but what kind of an aether it was—i.e. was it a mechanically explicable elastic solid aether?—was less clear. Maxwell did not concern himself with this question to any great extent, but rather attempted to formulate explanations of electrical, magnetic, and optical phenomena on the basis of the theory itself. In 1873 he re-presented his views in the monumental *Treatise on Electricity and Magnetism*. It is not necessary to go into this work in any detail except to point out the expressions for the energy of the electromagnetic field which he gives there. Though these are essentially the same as in the 1865 paper, the references of Maxwell's followers with whom we shall be concerned are usually to the *Treatise*.

In the sections 630–8 Maxwell investigated the distribution of energies in the field, and concluded:

The energy of the field therefore consists of two parts only, the electrostatic or potential energy:

$$W = \tfrac{1}{2} \int\int\int (Pf + Qg + Rh)\, dx\, dy\, dz \tag{5.1}$$

[where P, Q, and R are the components of the electromotive force intensity and f, g, h those of the electric displacement], and the electromagnetic or kinetic energy:

$$T = \frac{1}{8\pi} \int\int\int (a\alpha + b\beta + c\gamma)\, dx\, dy\, dz \tag{5.2}$$

[in which a, b, and c represent the components of the magnetic induction and α, β, and γ those of the magnetic force.]

Later in the *Treatise* Maxwell reasserted his commitment to the aether, citing again the connection between the optical aether and the electromagnetic aether:

In the theory of electricity and magnetism adopted in this treatise, two forms of energy are recognized, the electrostatic and the electrokinetic . . . , and these are supposed to have their seat, not merely in the electrified or magnetized bodies, but in every part of the surrounding space, where electric or magnetic force is observed to act. Hence our theory agrees with the undulatory theory in assuming the existence of a medium which is capable of becoming a receptacle of two forms of energy.

3. Fitzgerald's Electromagnetic Interpretation of MacCullagh's Aether

G. F. Fitzgerald represents a fairly typical Maxwell follower of the late nineteenth century. On the one hand, he was convinced of the worth of Maxwell's theory as he found it presented in the 1873 *Treatise*, though there were, of course, rough patches to smooth over and undeveloped areas in which to apply the theory. On the other hand, Fitzgerald also exhibits a belief fairly widespread during this period that the discovery of the mechanical aether theoretical basis of Maxwell's theory would constitute a significant step in the advancement of the theory.† Working in the "Dublin tradition" of optics and aether theory Fitzgerald took over MacCullagh's aether theory and the application of that theory to reflection and refraction, and showed how it could be used within Maxwell's theory. Maxwell had not been able to work out an explanation of reflection and refraction on the basis of his electromagnetic theory, as he was not able to satisfy himself as to the boundary conditions which should hold in his theory. Such an explanation was first presented by H. A. Lorentz in his doctoral dissertation in 1875, and I shall comment on this in the next chapter. Fitzgerald, not knowing of Lorentz' work, developed his own electromagnetic account of reflection and refraction in 1878. Maxwell refereed Fitzgerald's paper embodying this theory for the *Philosophical Transactions of the Royal Society* and commented

† See Glazebrook's (1885) comments on deficiencies in Maxwell's theory, and the letter by Heaviside to Hertz quoted below, on p. 90.

favorably on it, noting that though Lorentz had anticipated Fitzgerald on a number of points, the paper made several new contributions to the electromagnetic theory, particularly as regards reflection and refraction in magnetized media.[†]

Fitzgerald's method of attacking the problem, to characterize it in very broad terms, is (1) to use Maxwell's expressions for the kinetic and potential energies of the medium, then (2) to map the basic quantities of the electromagnetic theory into MacCullagh's aether and show that there is a parallelism in the energy equations of Maxwell's and MacCullagh's media, and finally, (3) to obtain both the equations of vibratory motion and the boundary conditions by using these energy expressions in Hamilton's Principle of Least Action.

Fitzgerald uses quaternion notation as well as Cartesian component notation in his essay and though I will not pretend to explain quaternions with any degree of adequacy, I will make some comments on the effective relation of them to the vector notation.

Following Maxwell, Fitzgerald defines the potential or electrostatic energy as:

$$W = -\tfrac{1}{2} \iiint S(E \cdot D)\, dx\, dy\, dz$$
$$= \tfrac{1}{2} \iiint (Pf + Qg + Rh)\, dx\, dy\, dz. \qquad (5.3)$$

The first expression is in quaternion notation, $S(E \cdot D)$ representing for our purposes the quaternion analogue of the negative of the vector dot product.[‡] E and D represent, of course, the electromotive force and electric displacement respectively. E and D are

[†] See the report by Maxwell on Fitzgerald's (1880) essay which is part of the Joseph Larmor Collection, Anderson Room, Cambridge University Library.
[‡] See Bork (1966b) for a concise discussion of the quaternion notation and some disputes that arose over its use.

understood to be related by:

$$E = \phi D \tag{5.4}$$

where ϕ is a dielectric function of the medium, not necessarily isotropic. The kinetic energy is as in Maxwell's writings given by:

$$T = -\frac{1}{8\pi} \iiint S(B \cdot H)\, dx\, dy\, dz$$
$$= \frac{1}{8\pi} \iiint (a\alpha + b\beta + c\gamma)\, dx\, dy\, dz \tag{5.5}$$

where B is the magnetic induction, H the magnetic force, and ae usual, $B = \mu H$ where μ is the coefficient of magnetic inductivs capacity.

Fitzgerald also makes explicit use of one of Maxwell's "curl equations", to use later parlance, which Fitzgerald writes as:

$$4\pi \dot{D} = V \nabla H \tag{5.6}$$

which is the same, for our purposes, as:

$$4\pi \frac{\partial \boldsymbol{D}}{\partial t} = \nabla \times \boldsymbol{H}. \tag{5.7}$$

At this point in his discussion, Fitzgerald introduces MacCullagh's aether. He does this by defining what amounts to an aether displacement vector R, which is MacCullagh's ξ, η, ζ, such that:

$$R = \int H dt, \tag{5.8}$$

which is equivalent, differentiating both sides with respect to time, to:

$$\dot{R} = H. \tag{5.9}$$

Substituting this definition in (5.6) or (5.7) above gives:

$$4\pi D = V \nabla R \tag{5.10}$$

if we integrate both sides of the result in the substitution in order to eliminate the differentiation with respect to time. In component notation (5.10) becomes:

$$4\pi f = \frac{d\zeta}{dy} - \frac{d\eta}{dz}, \quad 4\pi g = \frac{d\xi}{dz} - \frac{d\zeta}{dx}, \quad 4\pi h = \frac{d\eta}{dx} - \frac{d\xi}{dy}. \quad (5.11)$$

Either (5.10) or (5.11) tells us that Maxwell's dielectric displacement is the rotation of this "elastic solid" aether, and from (5.9) we can see that the magnetic force is the velocity of an aether stream in this aether.

Substitution of the R term into the Maxwell expressions for T and W reveals a striking analogy with the MacCullagh aether with respect to the form of the potential energy of the medium. In MacCullagh's aether we found that:

$$V = -\tfrac{1}{2}(a^2X^2 + b^2Y^2 + c^2Z^2). \tag{4.38}$$

If we compare (5.11) and MacCullagh's definitions of X, Y, and Z (see (4.37)) we see that the rotations are taken in an opposite sense, thus accounting for the difference in sign. The expressions then will differ by constant factors which can be adjusted so that the expressions (4.38) and (5.1) are equivalent.

Fitzgerald's actual expression for W in his version of MacCullagh's aether is obtained by substitution of (5.4) and (5.10) in (5.3) which yields:

$$W = -\frac{1}{32\pi^2} \int\int\int S(V\triangledown R \cdot \phi V \triangledown R) \, dx \, dy \, dz. \tag{5.12}$$

But Fitzgerald had to generalize MacCullagh's medium before he could build an analogue of the magnetic force into it. This was done, as suggested above, by the introduction of the streaming motion into the aether. The energy of this stream is, of course, kinetic, being energy of motion. Substitution of (5.9) into (5.2) yields:

$$T = \frac{1}{8\pi} \int\int\int \mu \dot{R}^2 \, dx \, dy \, dz \tag{5.13}$$

which is an acceptable expression for the mechanical kinetic energy of the aether if μ can be taken as a measure of the aether's inertial resistance.[†] (Fitzgerald's actual expression for (5.13) has a minus sign because of quaternion conventions.)

Having obtained expressions for the potential and kinetic energy of his aethereal medium, Fitzgerald applied the Hamilton formulation of the Principle of Least Action:

$$\delta \int (T-W)\,dt = 0 \qquad (5.14)$$

which was given in essentially the same form as (4.10) in the previous chapter. Substitution of (5.12) and (5.13) into (5.14) gives a complicated expression which can then be developed by standard mechanical methods, such as were followed by Green and MacCullagh in their earlier aether theories. Integration by parts gives rise, as usual, to two sets of integrals: (a) triple or volume integrals, which Fitzgerald refers to as "general integrals", and (b) double or surface integrals which Fitzgerald calls "superficial integrals".

The reader should already be able to see the similarity to the Green approach considered in some detail in Chapter IV. As with Green, and also with MacCullagh, the "general" integrals will yield the equation of motion of any disturbance, and can easily be made to give the equation of a plane wave by imposing the appropriate restrictions. The "superficial" integrals give the boundary conditions for reflection and refraction which hold at an interface between any two media, subject of course to the limitations of the assumptions made at the very beginning of Fitzgerald's paper which limited his inquiry to nonconductors with an isotropic μ.

By following MacCullagh's (1839) analysis fairly closely, Fitz-

[†] Whittaker (1960) presents Fitzgerald's theory in a somewhat different manner than what I have done here, my treatment being closer to Fitzgerald's original approach.

gerald was able to obtain the law of reflection, Snell's law of refraction, and the Fresnel sine and tangent laws exactly.

I have only included the first part of Fitzgerald's paper in which he introduces his generalization of the MacCullagh aether, and not that part in which he obtains the laws of reflection, refraction, and Fresnel's laws. No new principles are introduced here, and as the mathematics is carried through mostly in the quaternion form, with translations made to the Cartesian form from time to time, it would be difficult for the average reader to follow.

The reason for the introduction of the Fitzgerald analysis is to show one example in which the optical aether and the electromagnetic aether were brought together. The Fitzgerald analysis was later explored in a more detailed manner by Joseph Larmor, to whom I shall turn in a moment.[†]

It would, however, be somewhat misleading to argue that analyses like Fitzgerald's were completely satisfactory. We have already seen that the MacCullagh aether is deficient in an important dynamical way, so that even if an adequate reduction of the Maxwell theory to the generalized MacCullagh aether could be carried out, it would not constitute a reduction of the electromagnetic theory to mechanics. Larmor puzzled over this deficiency in the MacCullagh aether and eventually came to terms with it in a most interesting way as we shall see below. But to a growing number of late nineteenth-century physicists it was not clear that these rotational aethers were much better than interesting analogies.

Oliver Heaviside, who made a number of contributions to the development of Maxwell's theory, was one such person. Heaviside, though he lived until 1925, never gave up his belief in an aether and after Einstein's special theory of relativity had been accepted by most physicists, continued to criticize it as being too "abstract" and as wanting an aether. Earlier, in a letter to Hertz dated 13 September, 1889,[‡] Heaviside displayed the widespread feeling

[†] Sommerfeld (1892) also considered an aether similar to Fitzgerald's.
[‡] This letter is located in the Deutsches Museum, Munich.

which I mentioned above that Maxwell's theory required a clearer aether theoretical basis. Heaviside wrote:

> I believe it quite possible to frame a mechanical theory of a compressible aether which should lead to Maxwell's equations. But no doubt Maxwell's theory of displacement and induction in ether must remain (in spite of your and similar experiments to come) a Paper-Theory—as long as we do not know *what* functions of the ether D and B are! . . .

Later in the same letter Heaviside foreshadowed an aether yet to be developed by Joseph Larmor, writing:

> It often occurs to me that we may be all wrong in thinking of the ether as a kind of matter (elastic solid for instance) accounting for its properties by those of the matter in bulk with which we are acquainted; and that the true way could we only see how to do it, is to explain matter in terms of the ether, going from the simpler to the more complex.

Two years later Heaviside wrote up a short paper on an aether which was essentially the same as Fitzgerald's generalization of MacCullagh's aether. Heaviside, however, thinks of it more in line with a generalization of Kelvin's quasi-labile, sometimes called quasi-rigid, aether which I discussed in the last chapter.[†] The mode of presentation of this aether by Heaviside is also quite different from the more abstract work of Fitzgerald. Heaviside's approach, rather than proceeding through energy expressions and the Principle of Least Action, utilizes Newton's law of motion for translation and for rotation: $F = ma$, and torque $= I\alpha$, as expressed in their most elementary formulations.

Though the Heaviside rotational aether is not very important from the point of view of the history of aether theory, it does afford a different approach to the electromagnetic aether, and also points out how it may be applied to practical electromagnetic problems, such as telegraphy. It also shows in what ways the parallelism between such an aether and Maxwell's electromagnetic

† In 1890 Kelvin had applied his aether to magnetism. See his collected papers, vol. iii, p. 465.

theory begins to break down.† It is for these reasons that I have included it in the selections.

We may now move to consider the most highly developed optical-mechanical-electromagnetic aether theory which came out of the nineteenth century. This was Joseph Larmor's aether, which Whittaker suggested, I believe erroneously, was able "to withstand ... criticisms based on the principle of relativity, which shattered practically all rival concepts of the aether".

4. Joseph Larmor's Aether and the Electron

Larmor is perhaps not so well known today as the previously cited writers on the electromagnetic aether. Nevertheless, it was Larmor who not only brought the mechanical-electromagnetic aether to its most developed state, but who also was the first person (1897) to incorporate the Lorentz–Fitzgerald contraction within a *general* electromagnetic explanation of aberration phenomena. I will discuss the first of Larmor's contributions in this chapter, as it forms a natural conclusion to the work of Maxwell and Fitzgerald. For reasons which will become clearer later, Larmor's account of aberration best belongs in the next chapter after a preliminary discussion of Lorentz' work.

Larmor was trained at Queens College, Belfast, and at St. John's College, Cambridge, where he was first wrangler in 1880, followed by J. J. Thompson. After three years as professor at Queen's College, Galway, he returned to Cambridge, first as a lecturer. He assumed G. G. Stokes' Lucasian Professorship in 1903 which he held until he retired in 1932. He spent the remaining ten years of his life in Holywood near Belfast. In Larmor's obituary, A. E. Eddington (1942) discusses Larmor's strong feelings for the Irish

---

† This form of the rotational aether also had doubt cast on it by the experiment of O. Lodge (1897) which showed that a strong magnetic field would not influence the velocity of light, as one would expect to be the case if, as in this type of aether, a magnetic field is represented by an aether stream.

nation and countryside, and suggests that "it is no accident that *Aether and Matter* [Larmor's major work] is so largely a development of his countrymen MacCullagh, Hamilton, [and] Fitzgerald".

For whatever the reason, Larmor was a most vigorous proponent of MacCullagh's aether as understood through Fitzgerald's electromagnetic interpretation and Hamilton's Principle of Least Action. Larmor's inquiry into this aether began in 1893 and developed through the next five years. He republished a revised version of his essays in book form as *Aether and Matter* in 1900 for which he was awarded the Adam's Prize.

Larmor from the inception (1893) of his inquiry had felt that "our notions of what constitute electric and magnetic phenomena are of the vaguest as compared with our ideas of what constitutes radiation" and he believed that "many obstacles may be removed by beginning at the other end, by explaining electric actions on the basis of a mechanical theory of radiation, instead of radiation on the basis of electric actions".

Larmor's specific approach was, as he himself acknowledged, identical with that of Fitzgerald as regards the free aether. He used Fitzgerald's expressions for the potential and kinetic energy of the media which could variously be interpreted in either Maxwell's aether or MacCullagh's generalized aether. Larmor employed the same transformations as Fitzgerald to obtain a relation between the basic quantities of the "two" aethers, identifying electric displacement with aether rotation and magnetic force with aether velocity. The Principle of Least Action was also used to obtain the equations of motion of a disturbance propagated through the aether.

What is interesting and new about Larmor's analyses, other than the introduction of the electron which I shall discuss later, is the blend of critical self awareness about the problem of mechanical explanation together with an almost unshakable faith in the applicability of Hamilton's Principle to the aether. Larmor's position on the relation between mechanics and MacCullagh's aether apparently underwent some evolution between 1893 and 1900—an evolution which is worth discussing not only in connection with the evolution of the concept of the aether, but also because of certain parallels which it exhibits with the development of an electromagnetic view of nature, or electromagnetic "Weltbild", on the Continent.

I cited Stokes' criticism of MacCullagh's aether in the previous chapter. Larmor was well aware of these difficulties and in 1893 in his first paper on MacCullagh's aether Larmor wrote:

> As regards the rotational elasticity of this hydrodynamical aether on which we have made all radiative and electrical phenomena depend, it was objected in 1862 by Sir George Stokes to MacCullagh's aether, that a medium of that kind would leave unbalanced the tangential surface traction on an element of volume, and therefore could not be in internal equilibrium; and this objection has usually been recognized, and has led to MacCullagh's theory of light being put aside, at any rate in this country.

Larmor's (1893) reply to Stokes' criticism was rather unsatisfactory. It depended on conceiving of gravitation as a non-aethereal process which would provide the missing restoring force. Larmor himself called this use of gravitation a "saving hypothesis" and a "useful *deus ex Machina*", and he did not employ it again in his later analysis in *Aether and Matter*.†

In *Aether and Matter*, rather, instead of attempting to reduce electromagnetism to mechanics via the mechanical aether, Larmor suggested, as Heaviside had before him, that the mechanics of matter might be reduced to the actions of the electromagnetic aether, the latter conceived of as a kind of ultra-primitive matter or, if I may use the term, an *Ur*-aether. Larmor did not have a very clear idea of this *Ur*-aether, though in several places he

---

† See the letter from Larmor to Heaviside, 12 October, 1893, in the Heaviside Collection at the Institute for Electrical Engineers, London. A. Sommerfeld (1950) suggested that MacCullagh's aether involved "a 'quasielastic' body . . . *responsive to rotations relative to absolute space*"!

suggests that the actions of ordinary matter might be explained on its basis. For example, he indicates that the inertia of a particle of ordinary matter might well be understood in terms of the aether as characterized by Maxwell's equations or the Fitzgerald generalization of MacCullagh's theory. Elsewhere in the book he says: "An aether of the present type can hardly on any scheme be other than a medium, or mental construction if that term is preferred, prior to matter and therefore not expressible in terms of matter." In a long obituary which Larmor (1908) wrote on Lord Kelvin, Larmor discussed the Kelvin aether model which was presented in the previous chapter. His comments illuminate both Kelvin's model as well as his own views about the *Ur*-aether:

> It has come to pass that by making a model, with ordinary matter, of an elastic medium that has not the properties of ordinary matter, Lord Kelvin has vindicated to many minds if not entirely to his own, the power and cogency of mathematical analysis which can reach away without effort from the actual to the theoretically possible, and for example, make a mental picture of an aether which is not matter for the simple reason that it is something antecedent to matter.

It is essential, then, to distinguish in the selections I have extracted from *Aether and Matter* between a reduction to "dynamics" in Larmor's sense of the term, which is nothing more than the characterization of this *Ur*-aether with the aid of the Principle of Least Action, and a different kind of aether approach, such as we encountered in Green's theory. This latter approach, according to Larmor, "virtually identifies aether with a species of [ordinary] matter". Such an approach has led to difficulties which seem insoluble, but such difficulties can at least be "deferred" in the case of his own aether, Larmor maintains:

> if we are willing to admit without explanation the scheme of equations derived [in the first of the appended Larmor selections] from the form of energy functions for the aether, supposed stagnant, which is then postulated, in combination with the Principle of Least Action, and as a corollary, with an atomic structure of matter, involving electrons in its specification.

The aether had, during the seven years in which Larmor developed his theory, become very "aethereal" indeed, and Larmor, relinquishing the search for an ordinary dynamical foundation for his aether, seems to have moved in the same direction as those physicists who were developing an electromagnetic foundation of nature. Such physicists, among them W. Wien and M. Abraham, argued that since the "mass" of an electron could be understood in terms of the self-action of the electron's field on its own charge, rather than attempting to reduce electromagnetics and optics to mechanics, perhaps mechanics ought to be reduced to an electromagnetic theory.[†]

We have seen that Larmor largely followed in MacCullagh's and Fitzgerald's footsteps. But Larmor carried the aether theory considerably beyond what he had found. The most original contribution of Larmor's aether theory in its initial stages was, according to Fitzgerald himself, the introduction of the vortex atom of Lord Kelvin into such an aether. Fitzgerald wrote to Heaviside on 8 February, 1895 candidly commenting on Larmor's recent work and noted: "Larmor... has made a decided advance in pointing out that M'Cullagh's medium may have a common irrotational flow without any stresses so that vortex rings might exist in it.... I anyway had not appreciated this before."[‡]

Ironically enough Larmor gave up the vortex atom idea almost immediately and in its place substituted the mobile electron. The electron, or natural unit of electricity, which had received its name several years before from G. Johnstone Stoney, was conceived by Larmor to be a singularity in his aether, or in his own words, "analogous to... a simple pole in the... theory of a function of a complex variable". The vortex atom had been an important contribution of Lord Kelvin who had based his thoughts on some work by Helmholtz. Kelvin had conceived of the possibility of

---

[†] See M. Jammer's (1961) book for a brief discussion of this trend in physics. R. McCormmach also has an excellent unpublished paper on this topic (see Preface, p. ix).

[‡] The Fitzgerald letter is at the Institute for Electrical Engineers, London, in the Heaviside Collection.

material atoms being permanent vortex rings in a structureless homogeneous, and frictionless medium.† In his first full paper (1894) on the MacCullagh aether—Larmor had published an abstract of his theory in 1893—Larmor believed he could explain Ampère's theory of permanent magnetism by conceiving of the vortex atoms in his aether as electric currents. However, in an Appendix to this paper, dated 13 August, 1894, he admitted that vortex rings would not account for the magnetism and instead was driven to assume the existence of permanent charges, or electrons, and to conceive of magnetic molecules as "a single positive or right handed electron and a single negative or left handed one revolving round each other". Larmor realized that such a system should radiate, but as experiments revealed no such radiation, he proposed that no radiation should be released except when the "steady motion" of such a system was disturbed. This was somewhat *ad hoc* at the time, though something like it was introduced later by Bohr in his quantum theory of the hydrogen atom.

Larmor's electrons are freely mobile aether singularities which move through the aether "much in the way that a knot slips along a rope". Larmor's electron theory that resulted from the introduction of the electron hypothesis, along with the required appropriate modifications of the Fitzgerald equations of the aether, is very much like the electron theory of H. A. Lorentz. If it can be said—and reservations were expressed about this—that Fitzgerald found a generalization of the optical aether which was adequate to take into account Maxwell's electromagnetic aether, it can also be said that Larmor discovered a means of absorbing within a "dynamical" aether theory, the electron theory of Lorentz. Larmor claims that his own electron theory was independent of Lorentz', about which more will be said in the next chapter, but it is clear that in certain ways Larmor knew of and built on Lorentz' contributions. The question of independence is rather unclear though, and Lar-

† See Whittaker (1960), I, pp. 293 ff., for a discussion of the vortex atom·

mor's claims which he made in a letter to Lord Kelvin, written 14 November, 1899, ought to be carefully considered. Larmor wrote commenting on his forthcoming book *Aether and Matter:*

> When I began this train of ideas in 1894 led by your gyrostatic irrotational aether, I did not of course know that Lorentz was working out the same thing on a more abstract basis. In fact it was a long time before I perceived we were on the same lines while most others have not perceived it yet: which is a tribute either to the difficulty of the subject or to our imperfect powers of exposition. Yet I hold that they are in the main the same: and I derive therefrom confidence in the general scheme.†

Because of the importance of the electron theory in the development of late aether theory and early relativity theory it will be worthwhile showing how Larmor revised the Fitzgerald equations to incorporate electrons. A detailed discussion of the equations of the electron theory and of some of the differences of this theory with Maxwell's theory will be presented in the next chapter.

Larmor, like Lorentz (1892a) before him, says he cannot analyze the motions of individual electrons but must deal with averages and differential elements of volume which contain a number of electrons and molecules. When electrons are included in such volume elements, the equations which heretofore were based on Maxwell's equations for the free aether must be altered, as the displacement vector D or (f, g, h) is no longer circuital. Larmor then argues by two different chains of reasoning that a quantity equal to the true current, i.e. $e\dot{x}$, $e\dot{y}$, and $e\dot{z}$ must be added to the displacement current $\dot{f}$, $\dot{\bar{g}}$, and $\dot{h}$ to obtain a circuital vector. The equation which Larmor obtains, then, is equivalent to Lorentz' modification of the Maxwell curl equation, from:

$$\operatorname{curl} H = \frac{\partial D}{\partial t} \qquad (5.15)$$

---

† This letter is part of the Larmor Collection in the Cambridge University Library.

to:

$$\operatorname{curl} \boldsymbol{H} = \frac{\partial \boldsymbol{D}}{\partial t} + \varrho \boldsymbol{v} \qquad (5.16)$$

which is perhaps more familiar to the reader.

I have included both of Larmor's arguments in the appended selections and the reader may follow them there. The remainder of Larmor's Chapter VI which follows this modification is not included. In it Larmor assesses the consequences of introducing the electrons into the theory. His analyses are quite complicated and require the introduction of several auxiliary potential functions in order to develop the Principle of Least Action for the aether with electrons. Suffice it to say that what Larmor obtains is the inclusion of additional energy terms in his earlier expressions. He uses the Hamilton Principle and obtains an expression for the force with which the aether acts on an electron—an expression which is analogous to the Lorentz force expression. Larmor also obtains another interesting result concerning the amount of strain which an electron produces on the aether. He then applies his theory to problems of electrical conduction, double refraction, and black body radiation.

Some of the contributions for which Larmor is currently remembered, such as the radiation formula for a moving charge and the "Larmor precession", will not be referred to in these pages. Larmor's important contributions to aberration theory and to a clarification of the puzzling null result of the Michelson–Morley experiment will be considered in the next chapter.

CHAPTER VI

LORENTZ' AETHER AND THE ELECTRON THEORY: THE ELECTRODYNAMICS OF MOVING BODIES

H. A. LORENTZ represents a watershed figure in the history of modern physics. On the one hand he was an important—some might say the most important—theorist working in the Maxwell field tradition. He developed a theory of electrons which for a short time before quantum theory looked as if it might develop into a theory of such general scope that it would seriously rival classical mechanics.† But the electron theory did belong in the classical, meaning non-quantum, tradition of physics. Lorentz lived until 1928, and after his initial distrust and skepticism concerning the quantum theory, made several contributions to it.

1. The Development of Lorentz' Ideas of the Stagnant Aether and the Electron

Hirosige (1962) notes that when Lorentz was an undergraduate, he "carefully studied Maxwell's papers on electromagnetism", and "at the same time... enthusiastically studied Fresnel's wave theory of light and was deeply impressed by its penetrating lucidity".‡ Lorentz also concerned himself with Helmholtz' electromagnetic theory which was a generalized theory based on both the continental

† There are no adequate references on this topic, with the exception of an unpublished paper by R. McCormmach (see Preface, p. ix).
‡ See also G. L. DeHaas-Lorentz' (1957) for indications of the influence of Fresnel and Maxwell on Lorentz.

"action at a distance" approach, and Maxwell's contiguous "field" or "aether" approach to electromagnetism. In 1875 Lorentz submitted his doctoral dissertation in which, proceeding from Helmholtz' orientation, he was able to formulate explanations of reflection and refraction of light that could be applied within Maxwell's theory. Maxwell noted the importance of Lorentz' work in this regard, and commented very favorably on it. Maxwell himself, as I pointed out earlier in connection with Fitzgerald's theory, had not been able to work out an explanation of reflection and refraction on the basis of his own theory.

Several years later Lorentz formulated an explanation for dispersion utilizing the Helmholtzian theory and involving the interaction of molecules and the aether, but this is something that was later superseded by his electron theory, and I must refer the reader to other authors for comments on this point.[†]

In 1886, after having spent a number of years working on molecular and kinetic problems in physics, Lorentz undertook a long analysis of aberration phenomena, the problem with which this book began. As noted above, Lorentz had been impressed by Fresnel's theory, and in an essay published in 1886 that was based on his recent research, Lorentz, for reasons which were discussed in Chapter III, sided with a generalized version of Fresnel's stagnant aether, with its partial aether drag, against Stokes' theory of complete aether drag. Lorentz did not attempt at this point, however, to give an electromagnetic basis for his aether theory.

Maxwell had not proposed any satisfactory theory of the electrodynamics of moving bodies, and the few paragraphs in his *Treatise* on this subject were incompatible with the Fresnel partial dragging coefficient. Other theorists working in the Maxwell tradition were acutely aware of the problems and difficulties involved. For example, Oliver Heaviside wrote to Hertz in 1889 commenting on the general state of research in electromagnetic theory and on certain

---

† See Hirosige (1962).

crucial problems. Heaviside noted: "...there is the vexed question of the motion of the ether. Does it move when "bodies" move through it, or does it remain at rest? We know that there is an ether; the question is therefore a legitimate physical question which must be answered."† Several months earlier Heaviside had written Hertz:

> about aberration . . . the question is to explain Fresnel's result (confirmed by Michelson) electromagnetically. I have worked out the theory of the effect of motion of a dielectric on a wave going through it in terms of Maxwell's theory, but it does not explain Fresnel's results.‡

Lorentz did not present such an electromagnetic theory until 1892. I shall discuss this theory below.

There was one major difficulty with the Fresnel stagnant aether—and a stagnant aether was the type which was absorbed into Lorentz' electron theory—which caused great difficulties for Lorentz. This was the 1887 negative result of the Michelson–Morley interferometer experiment. I mentioned in Chapter III, when this experiment was discussed, that we would later consider Lorentz' reaction to the repetition of the criticized 1881 experiment, and it is to this that we now turn.

In the years between 1887 and 1892 Lorentz often thought about aberration problems, electromagnetic theory, and the Michelson–Morley null result. By early 1892 Lorentz had found the proper modification of Maxwell's theory which would permit him to derive the Fresnel convection coefficient, but which would not as yet enable him to account for the Michelson–Morley experiment. In this 1892 theory Lorentz proposed two types of entities: movable electrons (he termed them "ions" at the time) and a stagnant or immobile aether. The electrons were extremely small charged particles filling all material bodies. They were not considered singularities in the aether field, as were Larmor's electrons, but rather had a finite radius and a particular charge density distribution chosen

† The letter dated 14 August 1889 is in the Deutsches Museum, Munich.
‡ This letter, dated 1 April, 1889, is also in the Deutsches Museum.

so as to eliminate discontinuities at their interface. Lorentz introduced these movable charges into Maxwell's equations by a simple addition of a charge density term and a velocity term. Like Fresnel's aether, Lorentz' aether was immobile. It was actually even less mobile than Fresnel's aether which admitted of a partial drag within the interior of moving transparent ponderable bodies. The "partial drag" in Lorentz' theory was accounted for in terms of an alteration of the dielectrical constant by polarization in the moving media, and not by an actual partial aether drag. Unlike Fitzgerald's and Larmor's aethers, Lorentz' aether was not a dynamical medium; rather it was a ghostly framework or absolute reference system for Maxwell's and Lorentz' equations—though it was not necessarily identical with Newton's absolute space (see the first Lorentz selection on this point). Lorentz' aether was also the "seat" of the dielectric displacement and magnetic force fields.

Three years after his first long paper on the theory of electrons appeared, Lorentz re-presented his ideas in slightly different form in his classic *Versuch einer Theorie der Elektrischen und Optischen Erscheinung in Bewegten Korpern*. The *Versuch*, and I shall subsequently refer to it, is both a simplification and a development of the 1892 theory. My first Lorentz selection is from the Introduction of the *Versuch*, in which the author discusses some of the ways in which this later version constitutes an advance over both his earlier work and other competing theories. He also presents some of the experimental foundations for the electron theory and the immobile aether.

2. Lorentz' Response to the Michelson–Morley Experiment

One of the serious difficulties which the 1892 theory faced was, as I noted above, the problem of the negative result of the Michelson–Morley experiment. That this was of considerable concern to Lorentz, even after he had worked out the electron theory, is clear from a letter which he sent to Lord Rayleigh on 18 August,

1892, describing his new results but lamenting his failure to account for the interferometer experiment. Lorentz wrote:

> Fresnel's hypothesis [of a stagnant aether] taken conjointly with his [partial dragging] coefficient $1-1/n^2$, would serve admirably to account for all the observed phenomena were it not for the interferential experiment of Mr. Michelson, which has, as you know, been repeated after I published my remarks on its original form, and which seems decidedly to contradict Fresnel's views. I am totally at a loss to clear away this contradiction, and yet I believe if we were to abandon Fresnel's theory, we should have no adequate theory at all, the conditions which Mr. Stokes has imposed on the movement of aether being irreconcilable to each other.
>
> Can there be some point in the theory of Mr Michelson's experiment which has as yet been overlooked?

Lorentz concludes the letter by mentioning that he has endeavored:

> to apply the electromagnetic theory to a body which moves through the ether without dragging this medium along with it; my paper is now under the press and I hope, in a few weeks, to be able to offer you a copy of it. Assuming an approach which may appear somewhat startling but which may, as I think, serve as a working hypothesis, I have found the right value $1-1/n^2$ for F[resnel]'s coefficient. I hope to apply to some other problems the equations obtained, as for [example to] Fizeau's experiment on the rotation of the plane of polarization by a pack of glass plates.

Lorentz' perplexity is clear. Several months after this letter was sent, however, Lorentz found a solution. In an essay published in the 26 November, 1892 issue of the Amsterdam Academy's papers, Lorentz (1892b) proposed essentially the same hypothesis which Fitzgerald had suggested about three years earlier, namely that the motion of a body through the aether causes shrinkage of the body in the direction of motion in the ratio of $1 : (1 + v^2/2c^2)$, neglecting terms of order higher than $(v/c)^2$. The way in which the hypothesis is formulated is, curiously enough, almost identical for Lorentz and Fitzgerald: both ascribe contraction to transformation properties of *intermolecular* forces acting in the same way as *electrical* forces were known to be influenced by motion through the aether.

Lorentz, however, came to his contraction hypothesis independently of Fitzgerald's work.[†]

Lorentz returned again to these matters in the *Versuch* in which he proved in a more formal manner that the required contraction effect would follow if the cited intermolecular force transformations were granted.[‡]

3. Lorentz' Theorem of Corresponding States and its Development

Even in the *Versuch* the role which the contraction hypothesis played began to become more complex. In Chapter V of the work Lorentz introduced the notion of a "local time" which he employed in the proof of his most important "theorem of corresponding states". This theorem, which is asserted in the *Versuch* only for experiments to the first order of v/c, claims that if the standard classical Maxwellian type transformations for the dielectric displacement and magnetic force vectors are made by referring them to moving systems, then the same equations of the electron theory will hold in the moving system, as did so in the rest system, *if the local time is used for the time of the moving system.*

Lorentz employed this theorem in the *Versuch* in connection with first-order aberration experiments, and he returned to it again and again in later works, developing it under the impact of new experiments and theoretical considerations. In the *Versuch* the theorem is not applied to the Michelson–Morley experiment, but in a paper which appeared in 1899, Lorentz not only simplified his theory and the proof of this theorem by applying both time and space transformations immediately—in the *Versuch* a spatial transformation was not used in establishing the theorem of corresponding states—Lorentz also introduced transformation equations holding for the second order of v/c as a way of speculating about

---

[†] See the essays by Bork (1966a) and Brush (1966).
[‡] See Lorentz (1895), sections 23 and 89–92.

the effect of introducing a dielectric in the path of a light ray in the Michelson interferometer experiment. Lorentz believed such an experiment, as had recently been proposed by Liénard, would have a negative result, and his new transformation equations, which are almost equivalent to the famous 1904 transformation equations, predict this. But the second-order equations suffered from the defect of having an undeterminable coefficient ε present in them, and Lorentz could, at that time, think of no way to determine it.

Lorentz was not the first physicist to so extend his own Theorem of Corresponding States to the second order of v/c, and to tie the Lorentz–Fitzgerald Contraction effect to this. In 1897 Larmor, basing his ideas in part on Lorentz' *Versuch*, was apparently the first. I shall discuss this below.

I have included the 1899 paper of Lorentz' since it re-presents in a very succinct fashion, and in a still more simplified manner, many of the 1892 and 1895 contributions to physics, especially as regards the aether and aberration problems. The 1899 article has often been overlooked in treatments concerning the evolution of ideas which are of significance for the demise of the aether and the development of relativity theory. The reasons are not hard to discover. Lorentz himself confessed in 1904 that the issues he had attempted to deal with in 1899 could be treated much more satisfactorily, and he did so in his 1904 paper. With the exception of the *direct* application of the first-order Lorentz transformations, the speculative and incomplete extension of these equations to the second order of v/c, and a rather "startling" suggestion that motion in the aether influences the value of mass, this paper is a re-presentation of ideas that were worked out in the *Versuch*. But the 1899 paper does afford an excellent and simplified overview of Lorentz' ideas and their application to aberration phenomena, and it is probably easier to understand than the more famous 1904 paper, because of the additional complicating hypotheses in that paper (see Lorentz, 1904a). The latter paper is easily accessible to the reader, should he wish to follow up the development of Lorentz' ideas.

I have already noted that Lorentz' theory is an aether theory in the sense that the aether constitutes the absolute reference frame for the equations of the Lorentz electron theory. When a ponderable body is at rest with respect to this framework, instruments attached to the body measure "universal" time, rather than "local" time, length measurements are "correct" rather than fore-shortened by the Lorentz–Fitzgerald contraction, charge density is at its maximum value, and a similar situation holds for other transformed quantities which Lorentz discusses in his paper. The aether, accordingly, exercises causal effects on measuring instruments, material bodies, which move with respect to it. Lorentz was rather skeptical concerning the possibility of mechanically characterizing the aether medium. Though he apparently always believed that the aether propagated the electrical and magnetic forces which acted on electrons, by 1895 he had given up the mechanical approach, in the sense of the "least action" analysis, of the field equations which he had used in 1892 (and which we saw Larmor still used in 1900). In 1895 Lorentz was content to found the electron theory on the fundamental equations listed further below.[†] Lorentz' views about mechanical aether theories about the time of his 1899 paper are well expressed at the conclusion of a series of lectures he gave in 1901–2 on this topic. After covering in detail most of the mechanical theories and models which I have discussed earlier in this book, (including MacCullagh's and Fitzgerald's theories), Lorentz concluded by taking a position which was very similar to Hertz' widely known views. Lorentz wrote (1901):

> In what precedes a description was given of some of the attempts which were made in order to account for various phenomena, and especially the electromagnetic ones, by means of speculations about the structure and the properties of the aether. To a certain extent these theories were successful, but it must be admitted that they give but little satisfaction. For they

[†] R. McCormmach has a good discussion about Lorentz and the mechanical approach in one of his forthcoming papers (see Preface, p. ix).

become more and more artificial the more cases are required to be explained in detail. Of late the mechanical explanations of what is going on in the aether were, in fact, driven more and more to the background. For many physicists the essential part of a theory consists in an exact, quantitative description of phenomena, such *e.g.* as is given us by Maxwell's equations.

But even if one adheres to this point of view, the mechanical analogies retain some of their value. They can aid us in thinking about the phenomena, and may suggest some ideas for new investigations.

4. Lorentz' 1899 Essay on the Electrodynamics of Moving Bodies

The fundamental equations of the Lorentz theory, which are rather like those of Maxwell's theory in its Hertz–Heaviside form, are given in Lorentz' 1899 paper as follows:

$$\text{div } \boldsymbol{d} = \varrho \qquad \text{(Ia)}$$

$$\text{div } \boldsymbol{H} = 0 \qquad \text{(IIa)}$$

$$\text{curl } \boldsymbol{H} = 4\pi\varrho\boldsymbol{v} + 4\pi \frac{\partial \boldsymbol{d}}{\partial t} \qquad \text{(IIIa)}$$

$$4\pi c^2 \text{ curl } \boldsymbol{d} = -\frac{\partial \boldsymbol{H}}{\partial t} \qquad \text{(IVa)}$$

$$\boldsymbol{F} = 4\pi c^2 \boldsymbol{d} + \boldsymbol{v} \times \boldsymbol{H} \qquad \text{(Va)}$$

in which $\boldsymbol{d}$ is the dielectric displacement, $\boldsymbol{H}$ the magnetic force, ϱ the density to which the ponderable matter is charged, $\boldsymbol{v}$ the velocity of this matter, and $\boldsymbol{F}$ the force acting on it per unit charge. c is the velocity of light in the aether, t is the universal time, and ϱ, outside of the electrons, is equal to 0.[†]

Lorentz applied these equations to a system of bodies having a common velocity of translation $\boldsymbol{p}$, or $\boldsymbol{p}_x$, in the positive x direction of a Cartesian coordinate system. From the point of view of the moving coordinate system the Lorentz electron theory equations

---

[†] Lorentz uses less readable German letters in his exposition, and also follows the 19th century convention of using V for the speed of light. I have tried to use contemporary symbols here.

in their (a) form become more complicated and are transformed by a standard "Galilean" velocity addition principle and a space transformation into the (b) form (see the second Lorentz selection). The curl equations, in this (b) form, contain an explicit velocity term, p_x, and the Lorentz force equation also has in it an additional $(p_x \times H)$ term.

Lorentz now introduced new independent variables for x, y, z, and t. These were:

$$x' = \frac{c}{\sqrt{(c^2 - p_x^2)}} x$$
$$y' = y$$
$$z' = z$$
$$t' = t - \frac{p_x}{(c^2 - p_x^2)} x$$

in which "the last of these is the time, reckoned from an instant that is not the same for all points of space, but depends on the place we wish to consider. We may call it the *local time*, to distinguish it from the *universal time t*." (Lorentz, 1899.)

Substituting these new variables and differentiating with respect to them in accordance with the rules of the calculus, and introducing transformations for the magnetic force and the dielectric displacement—not directly for the latter but rather in terms of a closely related quantity termed the "electric force", E—Lorentz obtained his (c) form of the fundamental equations. If charged particles do not move *within* (that is with respect to) the moving (primed) reference system, one obtains a simple form of the fundamental equations (in their (c) form) in the moving primed reference system:

$$\text{div}' \, \boldsymbol{d}' = \frac{\varrho}{k} \quad (= \varrho') \qquad \text{(Ic*)}$$

$$\text{div}' \, \boldsymbol{H}' = 0 \qquad \text{(IIc*)}$$

$$\text{curl}' \, \boldsymbol{H}' = \frac{k^2}{c^2} \frac{\partial \boldsymbol{E}'}{\partial t'} \qquad \text{(IIIc*)}$$

$$\text{curl}' \, \boldsymbol{E}' = -\frac{\partial \boldsymbol{H}'}{\partial t'} \qquad (\text{IVc*})$$

$$F_x = E'_x, \quad F_y = \frac{E'_y}{k}, \quad F_z = \frac{E'_z}{k} \qquad (\text{Vc*})$$

where $k = \dfrac{c}{\sqrt{(c^2 - p_x^2)}}$.

(It should be noted that Lorentz does not, with the exception of Vc^*, explicitly simplify his equations in the above manner. I have accordingly added the * to the equation numbers to distinguish them from Lorentz' numbers. It will be clear from his paper, and from my subsequent discussion, however, that Lorentz was concerned to obtain, as far as *experimental* determination went, an invariant transformation of the equations. I should also add that the charge density transformation is worked out in this paper subsequent to the statement of the (c) form of the equations.)

Lorentz applied the non-simplified form of the (c) equations to electro-*static* phenomena (where the c^* form of the above equations *do* hold), and introduced the charge density transformation, cited above, on the principle that corresponding volume elements should have equal charge. (Recall that a moving ponderable body is contracted in the x direction in accordance with the Lorentz–Fitzgerald effect.) In equation form the charge density transformation is given by $\varrho' = \varrho/k$. Lorentz then showed that as a consequence the only change due to the motion of the moving system is a change in the resultant perpendicular electric forces, as given by (Vc^*). Therefore, Lorentz wrote: "every electrostatic problem for a moving system may be reduced to a similar problem for a system at rest, only the dimension in the direction of translation must be slightly different in the two systems".

Lorentz then showed, by making a series of approximations designed to eliminate terms involving $(v/c)^2$, how the (c) equations might "be applied to optical phenomena". I shall not review these approximations, which are dealt with by Lorentz himself in the second selection on pp. 261–66. Lorentz' intention is more important, and this is to arrive at his theorem of corresponding states:

> If, in a body or a system of bodies, without a translation, a system of vibrations be given, in which the displacement of the ions [or electrons] and the components of E' and H' are certain functions of the coordinates and the time, then, if a translation be given to the system, there can exist vibrations, in which the displacements and the components of E' and H' are the same functions of the coordinates and the *local* time.

Lorentz also noted that:

> This is the theorem, to which I have been led in a much more troublesome way in my "Versuch einer Theorie, etc.", and by which most of the phenomena, belonging to the theory of aberration may be explained.

Lorentz concluded his essay by proposing tentative second-order transformations. It should be noted again that this was done in response to an as yet untried variant of the Michelson–Morley experiment, but that these second-order transformations were unsatisfactory because they contained an ε term, which was a function of $(v/c)^2$, but which was indeterminable at the time.

5. Joseph Larmor's Extension of Lorentz' Theorem to the Second Order of v/c

As I noted above, second-order transformations for the equations of the electron theory for the free aether had already been introduced in 1897 by Joseph Larmor, though Lorentz does not give any evidence of having been aware of this. This analysis, which Larmor represented in *Aether and Matter*, is based on Lorentz' *Versuch*, which Larmor read shortly after it had been published in 1895. Lorentz' theorem of corresponding states was extended in a rather complex stepwise manner by Larmor. First the transformations for magnetic induction and aethereal elastic displacement were introduced:

$$(a', b', c') = (a, b+4\pi vh, c-4\pi vg)$$
$$(f', g', h') = (f, g-vc/4\pi C^2, h+vb/4\pi C^2)$$

in which v is the velocity of the primed system moving through the aether and C is the velocity of light. Substitution of these transformations in his curl equations gave Larmor new equations with extra terms involving v, even after terms involving $(v/C)^3$ were dropped, which were then eliminated by substituting:

t' for $t - vx/C^2$, and then dt_1 for $dt' \varepsilon^{-\frac{1}{2}}$,

x_1 for $x \varepsilon^{\frac{1}{2}}$,

a_1 for $a' \varepsilon^{-\frac{1}{2}}$,

f_1 for $f' \varepsilon^{-\frac{1}{2}}$. $\left(\text{Here } \varepsilon^{-\frac{1}{2}} = \sqrt{(1-(v/C)^2)}.\right)$

If this is done "the system", in Larmor's words, "comes back to its original isotropic form for free aether". In *Aether and Matter* (1900) an additional multiplying factor of $\varepsilon^{-\frac{1}{2}}$ is introduced for (a', b', c') and (f', g', h') or better (a_1, b_1, c_1) and (f_1, g_1, h_1) to ensure that the values of the electrons remain invariant under the transformations.

If this later modification can be ignored for the moment, it is clear that Larmor was the first physicist to extend the Lorentz theorem to quantities involving the second-order quantities of v/c. That Larmor realized the significance of this extension is also clear from his 1897 statement that followed the above analysis. Larmor wrote:

> If this argument is valid, it will confirm the hypothesis of Fitzgerald and Lorentz [the contraction effect], to which they were led as the ultimate resource for the explanation of the negative result of Michelson's optical experiments... .

Larmor's reasoning for the lack of aberrational effects, and his derivation of the Fresnel convection coefficient and the Lorentz–Fitzgerald contraction are included in the Larmor selections in the present book. For reasons of ease and elegance of exposition, I decided to include selections from *Aether and Matter* rather than from his earlier separate papers. In this later (1900) work Larmor presented his theory of compensation effects—for this is what the transformations amount to—for the moving electromagnetic or optical system in two stages. First, transformations are introduced to eliminate *first* order aberrational effects. Then, in another chapter, the theory of compensation effects and the transformations are extended to include *second* order aberrational effects. This is as far as Larmor took his aberration theory, however, and he never seemed inclined, before relativity theory, toward extending the compensation effects to include all orders of the quantity (v/c). Lorentz' approach, however, was somewhat different.

6. Lorentz' Later Papers on Electrodynamics of Moving Bodies

Lorentz did not cease his labors on the electron theory with his 1899 paper but soon went on to apply it in later papers to the Zeeman effect and other electromagnetic and optical phenomena which I cannot discuss in these pages. In 1904 he published (Lorentz, 1904b) an influential encyclopaedia article on the electron theory, and his famous paper (Lorentz, 1904a), which is often cited as an important precursor of Einstein's relativistic theory, also appeared in this year.

In the latter (1904a) paper Lorentz found a means of computing the value of the ε term which he had not been able to determine in 1899. By this time Lorentz also found it necessary to react to both Poincaré's criticism that a new hypothesis seemed to be required every time a new experiment associated with the electron theory was performed, and also to account for the unexpectedly negative results of the Trouton–Noble and Rayleigh–Brace experiments. I cannot discuss these experiments in the space available here; suffice it to say that they required a thoroughgoing revision of Lorentz' original justification for the contraction hypothesis, for in order to attain his goal of articulating a *generalized* theorem of corresponding states—for *many* electromagnetic phenomena to *all* orders of v/c—Lorentz had to postulate the contraction effect at the level of the individual electron. No longer would it be *sufficient* to ascribe it to the transformational properties of intermolecular forces, though this might figure in an explanation of a macroscopic contraction effect. (See my (1969) for more details.)

I shall close with some brief observations about the relation of the generalized theorem of corresponding states to Einstein's relativity principle.

Lorentz—certainly this holds for Larmor as well—is categorically a non-relativist in his interpretation of the transformations which bear his name. His understanding of them can be determined from a careful reading of his papers, and especially from his 1906 lectures published in 1909 as the *Theory of Electrons*. For Lorentz, an *inverse* spatial transformation can produce a spatial *dilation* effect: for example, he says that "if the electrons in S [the moving system] are spheres, those in S_0 [the rest system] are lengthened ellipsoids". Such an interpretation is also supported by Lorentz' formalism for inverse transformations, which simply uses reciprocals. Such dilations are clearly non-Einsteinian, for whom there were reciprocal contraction effects. (I would suggest, parenthetically, that Poincaré also seems to have this Lorentzian interpretation in his "relativistic" theory, because of one of the tasks of the "Poincaré pressure", but I cannot go into this here.)

The Lorentz aether, then, is important in the sense that it is positional: it serves as a unique reference system for Maxwell's and Lorentz' equations, and it exerts causal effects on charged objects that move through it. It is apparently also, in some unknown way, the bearer of the d and H fields. In Lorentz' theory of the aether we have, then, a dematerialized aether, largely divorced from the aether's usual task of serving as a dynamical basis for the field equations. Lorentz' aether still possessed certain "substance"-like properties, however. That this is so is supportable as we shall see by referring to Lorentz' 1909 monograph *The Theory of Electrons*.

After learning of and reflecting on Einstein's theory of relativity and on Einstein's independent and radically different derivation of transformation equations which were very much like his 1904 equations,[†] Lorentz re-analyzed the relation between moving frames of reference and considered what data two observers, one stationary with respect to the aether, A_0, and another observer, A, who moves through the aether would obtain from measurements within their own reference systems and between their reference systems. Lorentz concluded that:

> It will be clear by what has been said that the impressions received by the two observers A_0 and A would be alike in all respects. It would be impossible to decide which of them moves or stands still with respect to the ether, and there would be no reason for preferring the times and lengths measured by the one to those determined by the other, nor for saying that either of them is in possession of the "true" times or the "true" lengths. This is a point which Einstein has laid particular stress on, in a theory in which he starts from what he calls the principle of relativity, i.e. the principle that the equations by means of which physical phenomena may be described are not altered in form when we change the axes of coordinates for others having a uniform motion of translation relatively to the original system.

But Lorentz went on to register his disagreement with Einstein and to defend, rather weakly and metaphysically in the opinion of this author, his own approach. Lorentz wrote:

> I cannot speak here of the many highly interesting applications which Einstein has made of this principle. His results concerning electromagnetic and optical phenomena (leading to the same contradiction with Kaufmann's results that was pointed out in § 179) agree in the main with those which we have obtained in the preceding pages, the chief difference being that Einstein simply postulates what we have deduced, with some difficulty and not altogether satisfactorily, from the fundamental equations of the electromagnetic field. By doing so, he may certainly take credit for making us see

---

[†] It is unfortunately impossible in this book to show how Einstein, by approaching the problems from an entirely different point of view, cut through so much of the complexity of late-nineteenth-century electrodynamics. His approach involves a new kinematics: a new understanding of time and simultaneity, and of the relations of space and time, that discloses the true significance of Lorentz' transformations. See my (1970) for more details.

in the negative result of experiments like those of Michelson, Rayleigh and Brace, not a fortuitous compensation of opposing effects, but the manifestation of a general and fundamental principle.

Yet, I think, something may also be claimed in favour of the form in which I have presented the theory. I cannot but regard the ether, which can be the seat of an electromagnetic field with its energy and its vibrations, as endowed with a certain degree of substantiality, however different it may be from all ordinary matter. In this line of thought it seems natural not to assume at starting that it can never make any difference whether a body moves through the ether or not, and to measure distances and lengths of time by means of rods and clocks having a fixed position relatively to the ether.

It would be unjust not to add that, besides the fascinating boldness of its starting point, Einstein's theory has another marked advantage over mine. Whereas I have not been able to obtain for the equations referred to moving axes *exactly* the same form as for those which apply to a stationary system, Einstein has accomplished this by means of a system of new variables slightly different from those which I have introduced. I have not availed myself of his substitutions, only because the formulae are rather complicated and look somewhat artificial, unless one deduces them from the principle of relativity itself.

To close, I would also call attention to the fact that Lorentz believed, until 1914, that experimental phenomena might be discovered that would distinguish an absolute aether frame of reference. He was, as he said in his *Theory of Electrons*, "startled", and not entirely sympathetic, to Einstein's interpretation of the transformation equations. Later writings, for example Lorentz' 1914 Haarlem lectures and the 1915 edition of the *Theory of Electrons*, indicate that Lorentz was apparently persuaded of the worth of the relativistic approach, even if he was not entirely converted.

7. Einstein, Relativity, and the Elimination of the Aether

It should be clear from the preceding section that the Lorentz transformations would not require the elimination of the aether if they could be legitimately interpreted in some asymmetrical manner. Their original interpretation was non-reciprocal, but to-be consistent with experimental results this non-reciprocal interpretation had to be forsaken.† As soon as there is complete reciprocity between all reference frames, the aether becomes physically vacuous and vestigial. In the long quote above, Lorentz admitted in effect that this would be so. Since most physicists, including Lorentz himself, had by this time accepted the field intensities d and H as realities in their own right, and not as manifestations of a rotation or a velocity stream of aether, the aethereal substratum no longer possessed either positional or dynamical reasons for existence.

What physicists were left with as a consequence of Maxwell's, Lorentz', and Einstein's work was no aether at all. The tasks of the nineteenth-century aether are currently fulfilled by "fields" which transform in complete accord with the reciprocity interpretation of the Lorentz transformations. The only remanents of the elastic solid aether in contemporary physics are purely formal analogies which are embodied in the linear field equations.

Other aethers have occasionally appeared during the twentieth century. Einstein once suggested that the *general* theory of relativity had a type of aether associated with it, and P. A. M. Dirac proposed not very long ago that the sea of negative energy state electrons which filled empty space might well be considered as an aether. In the preface to his monograph on the history of the aether, Whittaker (1960) also argued on quantum electrodynamical considerations that the aether was a viable concept:

> . . . with the development of quantum electrodynamics, the vacuum has come to be regarded as the seat of the "zero-point" oscillations of the electromagnetic field, of the "zero-point" fluctuations of electric charge and current, and of a "polarization" corresponding to a dielectric constant different from unity. It seems absurd to retain the name "vacuum" for an entity so rich in physical properties, and the historical word "aether" may fitly be retained.

† It is not even clear to me that a formally consistent asymmetrical interpretation can be given to the Lorentz transformations in their Einstein form, i.e. with correct charge density and velocity addition expressions, because of the fact that they form a group.

The legitimacy of calling these new entities of Einstein, Dirac, and Whittaker "aether" is debatable; what is clear, however, is that the nineteenth-century aether has been relegated to that Ideal Realm populated by Caloric, Phlogiston, Epicycles, and other scientific concepts that have done their work so well as to have forced science beyond them.

REFERENCES

ARAGO, F. (1830) *Comptes Rendus* **8**, 326. (The experimental results were submitted to the French Institute in 1810.)
ARAGO, F. and FRESNEL, A. (1819) *Ann. de Chimie* **10**.
BASSET, A. B. (1892) *A Treatise on Physical Optics*, Deighton, Bell & Co., Cambridge.
BORK, A. (1966a) *Isis* **57**, 199.
BORK, A. (1966b) *Am. J. Phys.* **34**, 202.
BOUSSINESQ, J. (1867) *Comptes Rendus* **65**, 235.
BRACE, D. B. (1904) *Phil. Mag.* **7**, 317.
BRADLEY, J. (1728) *Phil. Trans. Roy. Soc.* **35**, 406.
BROMBERG, J. (1968) *Arch. Hist. Exact. Sci.* **4**, 218.
BRUSH, S. (1966) *Isis*, **58**, 230.
CAUCHY, A. (1839) *Comptes Rendus* **9**, 676, 726.
COTES, R. (1713) Preface to Newton's *Principia*, trans. A. MOTTE, rev. F. CAJORI. University of California Press Edition, 1962. Berkeley.
CREW, H. (ed.) (1900) *The Wave Theory of Light*, American Book Co., New York.
DEHAAS–LORENTZ, G. L. (ed.) (1957) *H. A. Lorentz: Impressions of His Life and Work*, North-Holland, Amsterdam, 15.
EDDINGTON, A. E. (1942) *Roy. Soc. Lond. Obit. Notices of Fellows* **4**, 197.
EHRENFEST, P. (1916) Published in *Collected Scientific Papers*, ed. M. KLEIN (1959), New York, Interscience, p. 471.
FEYERABEND, P. K. (1962) in *Minn. Stud. in Phil. of Sci.* **3**, ed. H. FEIGL and G. MAXWELL, University of Minnesota Press, Minneapolis, p. 28.
FEYERABEND, P. K. (1965a) in *Beyond the Edge of Certainty*, ed. R. COLODNY, Prentice-Hall, Englewood Cliffs, New Jersey.
FEYERABEND, P. K. (1965b) in *Boston Stud. in Phil. of Sci.* **2**, ed. R. S. COHEN and M. W. WARTOFSKY, Humanities Press, New York, p. 223.
FITZGERALD, G. F. (1880) *Phil. Trans. Roy. Soc.* **171**, 691 (reprinted as Paper 7).
FIZEAU, H. (1851) *Comptes Rendus* **33**, 349.
FRESNEL, A. (1818) *Ann. de Chimie* **9**, 57 (reprinted as Paper 1).
FRESNEL, A. (1821) Submitted to the Academy in 1821; published in *Oeuvres*, II, 261.
FRESNEL, A. (1826) *Mem. de l'Acad.* **5**, 339. (This is Fresnel's prize essay on diffraction crowned by the French Academy in 1819.)

REFERENCES

GLAZEBROOK, R. T. (1885) *Brit. Assoc. Reports* **55**, 157.
GLAZEBROOK, R. T. (1888) *Phil. Mag.* (5) **26**, 521.
GREEN, G. (1838) *Trans. Camb. Phil. Soc.* **7**, 1, 113 (reprinted as Paper 4).
GRÜNBAUM, A. (1963) *Philosophical Problems of Space and Time*, New York: A. Knopf.
HEAVISIDE, O. (1891) *Electrician*, **26**, 360. Also reprinted in his *Electromagnetic Theory*, Dover Edition (1950), p. 32 (reprinted as Paper 8).
HERTZ, H. (1888) *Ann. d. Phys.* **34**, 551, 610.
HERTZ, H. (1890) *Ann. d. Phys.* **40**, 577.
HESSE, M. B. (1965) *Forces and Fields*, Littlefield, Adams & Co., Totowa, New Jersey.
HIROSIGE, T. (1962) *Japan Stud. Hist. of Sci.* **1**, 101.
HIROSIGE, T. (1965) *Japan Stud. Hist. of Sci.* **4**, 117.
HIROSIGE, T. (1966) *Japan Stud. Hist. of Sci.* **5**, 1.
HOLTON, G. (1960) *Am. J. Phys.* **28**, 627.
HOLTON, G. (1964) in *Melanges Alexandre Koyre* **2**, Hermann, Paris.
HOOKE, R. (1665) *Micrographia*, Martyr & Allestry, London.
HUYGENS, C. (1690) *Traité de la lumière...*, Leyden (Engl. trans. by S. P. THOMPSON, Dover Edition, New York, 1962).
JAMIN, M. J. (1850) *Ann. de Chimie* (3) **29**, 263.
JAMMER, M. (1961) *Concepts of Mass in Classical and Modern Physics*, Harvard University Press, Cambridge.
KELVIN, Lord. See THOMSON, W. (Lord Kelvin).
KIRCHHOFF, G. (1876) *Abh. der Berl. Akad.*, 2 Abt., p. 57.
KUHN, T. S. (1962) *The Structure of Scientific Revolutions*, University of Chicago Press, Chicago.
LAGRANGE, L. (1788) *Méchanique Analytique*, Veuve Desaint, Paris.
LARMOR, J. (1893) *Proc. Roy. Sci.* **54**, 438.
LARMOR, J. (1894) *Phil. Trans. Roy. Soc.* **185**, 719.
LARMOR, J. (1895) *Phil. Trans. Roy. Soc.* **186**, 695.
LARMOR, J. (1897) *Phil. Trans. Roy. Soc.* **190**, 205.
LARMOR, J. (1900) *Aether and Matter*, Cambridge University Press, Cambridge (selections reprinted as Paper 9).
LARMOR, J. (1907) "Aether" in *Encycl. Brittanica*, 11th edition.
LARMOR, J. (1908) *Proc. Roy. Soc.* A, **81**, Appendix iii.
LINDSAY, R. B. and MARGENAU, H. (1957) *Foundations of Physics*, Dover Edition, New York.
LLOYD, H. (1834) *Brit. Assoc. Reports*, **4**, 295.
LODGE, O. (1897) *Phil. Trans. Roy. Soc.* **189**, 149.
LORENTZ, H. A. (1892a) *Arch. Néerl.* **25**, 363.
LORENTZ, H. A. (1892b) *Versl. Kon. Akad. Wetensch. Amsterdam* **1**, 74.
LORENTZ, H. A. (1895) *Versuch einer Theorie der Elektrischen und Optischen Erscheinungen in Bewegten Körpern*, Brill, Leyden (Introduction reprinted as Paper 10).

LORENTZ, H. A. (1899a) *Proc. Roy. Acad. Amsterdam*, **1**, 427 (reprinted as Paper 11).

LORENTZ, H. A. (1899b) *Proc. Roy. Acad. Amsterdam* **1**, 443.

LORENTZ, H. A. (1901) A set of lectures given in the academic year 1901–2 and published in 1927 as part of Vol. I of his *Lectures on Theoretical Physics*, trans. L. SILBERSTEIN and A. P. H. TRIVELLI, Macmillan, London.

LORENTZ, H. A. (1904a) *Proc. Roy. Acad. Amsterdam*, **6**, 809.

LORENTZ, H. A. (1904b) Article on "Electronentheorie" in *Encyklopädie der math. Wissenschaften*, 2, Leipzig.

LORENZ, L. V. (1861) *Pogg. Ann.* **114**, 238.

MACCULLAGH, J. (1848) *Trans. Roy. Irish Acad.* **21**, 17 (paper presented in 1839), (reprinted as Paper 5).

MACH, E. (1893) *The Science of Mechanics*, Open Court, LaSalle, Illinois.

MALUS, E. L. (1811) *Mém. presentés a l'Inst. por divers Savans*, **2**, 303.

MAXWELL, J. C. (1856) *Trans. Camb. Phil. Soc.* **10**, 27.

MAXWELL, J. C. (1961–2) *Phil. Mag.* **21**, 161, 281, 338; **23**, 12, 85.

MAXWELL, J. C. (1865) *Phil. Trans. Roy. Soc.* **65**, 459.

MAXWELL, J. C. (1873) *A Treatise on Electricity and Magnetism*, Dover Edition (1954), New York.

MICHELSON, A. A. and MORLEY, E. W. (1886) *Am. J. Sci.* (3), **31**, 377.

MICHELSON, A. A. and MORLEY, E. W. (1887) *Phil. Mag.* (5), **24**, 499; and *Am. J. Sci.* (3) **34**, 333 (reprinted as Paper 3).

NAGEL, E. (1961) *The Structure of Science*, Harcourt, Brace, and World, New York.

NEWMANN, F. (1837) *Abh. Berl. Ak. aus dem Jahre 1835, Math Klasse*, 1.

POPPER, K. R. (1959) *The Logic of Scientific Discovery*, Hutchinson, London.

PRESTON, T. (1895) *Theory of Light*, Macmillan, New York.

RAYLEIGH, Lord, See STRUTT, J.

ROSENFELD, L. (1956) *Nuovo Cimento, Supp.* (10), **4**, 1630.

SCHAFFNER, K. F. (1967) *Phil. Sci.* **34**, 137.

SCHAFFNER, K. F. (1969) *Am. J. Phys.* **37**, 498.

SCHAFFNER, K. F. (1970) *Minn. Stud. in Phil. of Sci.*, ed. R. Stuewer, University of Minnesota Press, Minneapolis.

SHANKLAND, R. S. (1964) *Am. J. Phys.* **32**, 16.

SOMMERFELD, A. (1892) *Ann. d. Phys.* **46**, 139.

SOMMERFELD, A. (1933) *Electrodynamics: Lectures on Theoretical Physics*, Vol. III, trans. E. G. RAMBERG. Based on lectures given in 1933–4. 1964 Edition by Academic Press, New York.

SOMMERFELD, A. (1950) *Mechanics of Deformable Bodies*, trans. G. KUERTI, Academic Press, New York. (Based on early lectures, similar to those cited in Sommerfeld, 1933.)

STOKES, G. G. (1845) *Phil. Mag.* **27**, 9 (reprinted as Paper 2). (See note on p. 136.)

STOKES, G. G. (1849) *Trans. Camb. Phil. Soc.* **9**, 1.

REFERENCES

STOKES, G. G. (1862) *Brit. Assoc. Reports* **32**, 253.
STRUTT, J. (1871) *Phil. Mag.* (4) **42**, 81.
STRUTT, J. (1892) *Nature* **45**, 499.
STRUTT, J. (1902) *Phil. Mag.* **4**, 215.
SWENSON, L. (1962) "The Ethereal Aether: a Descriptive History of the Michelson–Morley Aether Drift Experiments 1880–1930", Unpublished Ph. D. Dissertation, Claremont University.
THOMSON, W. (Lord Kelvin) (1888) *Phil. Mag.* (4) **26**, 414.
THOMSON, W. (Lord Kelvin) (1890) *Math. and Phys. Papers*, III, London, art. 100, 466 (reprinted as Paper 6).
THOMSON, W. (Lord Kelvin) (1884 and 1904) *Baltimore Lectures on Molecular Dynamics*, Johns Hopkins, Baltimore.
THOMSON, J. J. (1885) *Brit. Assoc. Reports* **55**, 971.
TRICKER, R. A. R. (1966) *The Contributions of Faraday and Maxwell to Electrical Science*, Pergamon, Oxford.
TROUTON, F. and NOBLE, H. R. (1903) *Phil. Trans. Roy. Soc.* A, **202**, 165.
VERDET, E. (1869) *Lecons d'Optique Physique*, **1**, Levistal, Paris.
WHITTAKER, E. T. (1960) *A History of the Theories of Aether and Electricity*, 2 vols., Harper & Bros., New York.
WILLIAMS, L. P. (1965) *Michael Faraday, A Biography*, Basic Books, New York.
YOUNG, T. (1800) *Phil. Trans. Roy. Soc.* **90**, 106.
YOUNG, T. (1802) *Phil. Trans. Roy. Soc.* **92**, 12, 387.
YOUNG, T. (1804) *Phil. Trans. Roy. Soc.* **94**, 1.
YOUNG, T. (1807) *Course of Lectures on Natural Philosophy*.
YOUNG, T. (1809) *Quarterly Review*, Nov. 1809.

PART 2

1. LETTER FROM AUGUSTIN FRESNEL TO FRANÇOIS ARAGO, ON THE INFLUENCE OF THE MOVEMENT OF THE EARTH ON SOME PHENOMENA OF OPTICS*†

My dear friend,

By your fine experiments on the light from the stars, you have shown that the movement of the terrestrial globe has no perceptible influence upon the refraction of rays emanating from these stars. Within the corpuscular theory, as you have pointed out, this remarkable result can only be explained by supposing that luminous bodies transmit to the particles of light an infinite number of different velocities, and that these particles only affect the organ of sight when travelling at one of these velocities, or at least between very close-set limits, so that an increase or decrease of a ten-thousandth part is more than enough to prevent their detection. The necessity for this hypothesis is not the least difficulty attaching to the corpuscular theory; for on what does vision depend? Upon the impact of the light particles on the optic nerve? In this case such an impact would not be rendered imperceptible by an increase

---

* *Ann. de Chimie* **9**, 57 (1818).

† Taken from a letter to Léonor Fresnel, 5 September 1818 (LIX): ". . . I have recently been engaged on a small work to which I attach some importance. I have proved that, supposing the earth to be sufficiently porous to the ether which penetrates and surrounds it, not to transmit to it more than a minute part of its velocity, not exceeding, for example, an hundredth part, one could explain satisfactorily not only the aberration of the stars, but also all the other optical phenomena which are complicated by the movement of the earth, etc." (H. de St.).

in velocity. Upon the way in which the particles are refracted within the pupil? But red particles, for example, whose velocity had been diminished even by a fiftieth part, would still be refracted less than violet rays, and would not leave the spectrum which defines the limits of vision.

You have enjoined me to examine whether the result of these observations could be reconciled more easily with the theory in which light is considered as being vibrations of a universal fluid. It is all the more necessary to find an explanation within this theory, because the theory applies to terrestrial objects; for the velocity of wave propagation is independent of the movement of the body from which the waves emanate.

If one were to admit that our earth transfers its movement to the ether surrounding it, it would be easy to see why the same prism would always refract light in the same way, whatever direction it came from. But it appears impossible to explain the aberration of stars by this hypothesis: I have been unable, up to the present at least, to understand this phenomenon clearly except by supposing that the ether passes freely through the globe, and that the velocity communicated to this subtle fluid is only a small proportion of the velocity of the earth, not exceeding, for example, an hundredth part.

However extraordinary this hypothesis may appear at first sight, it does not seem to me at all incompatible with the idea of the extreme porosity of bodies which the greatest physicists have arrived at. It may indeed be asked how, while a very thin opaque body is capable of intercepting light, a current of ether can pass through our globe. While not claiming to meet this objection completely, I shall nevertheless point out that the two kinds of movement are too unlike in character for observations made in connection with one to be applicable to the other. The movement of light is not a current, but a vibration of the ether. One may see how the small elementary waves into which light divides when passing through a body may, in certain cases, be out of phase on coming together again, by reason of the different paths they have taken or the different amounts by which they have been slowed down; this prevents the propagation of vibrations, or alters them in such a way as to remove their light-giving property, as occurs in a very striking manner with black bodies; while the same circumstances would not prevent the establishment of a current of ether. The transparency of hydrophane† is increased by wetting it, and it is evident that the interposition of water between its particles, while favouring the propagation of light vibrations, must on the contrary prove a small additional obstacle to the establishing of a current of ether. This is a good demonstration of the great difference which exists between the two types of movement.

The opacity of the earth is therefore not a sufficient reason to deny the existence of a current of ether between its molecules, and one may suppose it porous enough to communicate to this fluid only a very small part of its movement.

With the aid of this hypothesis the phenomenon of aberration is as easily explained by the theory of waves as by the corpuscular theory; for it arises out of the displacement of the optical instrument while the light is travelling through it. Now according to this hypothesis, the light waves do not participate to any perceptible extent in the movement of the telescope, which I am assuming to be pointed at the true location of the star, and the image of the star lags behind in relation to the cross-hair situated in the eyepiece of the telescope, by an amount corresponding to the distance covered by the earth while the light is travelling down the telescope.

It now remains to explain why, by the same hypothesis, the apparent refraction does not vary with the direction of the light rays relative to the movement of the earth.

---

† Hydrophane: a variety of opaque or partly translucent opal which absorbs water upon immersion and becomes transparent. [Translator's note.]

Let *EFG* (Fig. 1.1) be a prism, of which one side *EF* is placed at right angles both to the ecliptic and to the incident rays, which are thus travelling in the same direction as the earth: if the prism's movement has an influence upon their refraction, this is the case where it must be most apparent. I am supposing that the rays are moving in the same direction as the prism.

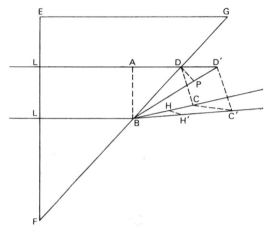

Fig. 1.1

Since they strike the surface of entry at right angles, the rays do not undergo refraction at this side of the prism, and it is only the effect produced by the second surface which needs to be considered. Let *LD* and *LB* be two of these rays, striking the surface of exit at points *D* and *B*. Let *BC* be the direction taken by the ray *LB* when it leaves the prism, in the case when the prism is stationary. If a perpendicular is dropped from the point *D* onto the emerging ray, and starting from point *B* a line *BA* is drawn perpendicular to the incident ray, the light must travel from *A* to *D* in the same time as from *B* to *C*: this is the law determining the direction of the refracted wave *DC*. But as the prism is being carried along by the movement of the earth while the light is travelling from *A* to *D*,

LETTER FROM FRESNEL TO ARAGO

point D is displaced, as a result of which the difference between the two paths travelled by the rays LD and LB is increased, which necessarily changes the angle of refraction. FG represents the surface of emergence; let D' be the point where, after the incident wave has reached AB, the ray AD strikes this surface and leaves the prism. Let BC' be the new direction of the refracted rays. The perpendicular $D'C'$ will represent the direction of the emerging wave, which must satisfy the general condition that AD' is travelled by the light in the same time as BC'. But in order to establish the relative lengths of these two intervals, it is necessary to calculate the variation introduced by the movement of the prism in the velocity of the light waves travelling through it.

If this prism carried along with it all of the ether it contains, the whole of the medium acting as a vehicle for the waves would thus participate in the movement of the earth, and the velocity of the light waves would be equal to their velocity in a supposedly stationary medium, plus the velocity of the earth. But the case in point is more complicated; it is only a part of the medium which is carried along by our earth—namely, the proportion by which its density exceeds that of the surrounding ether. By analogy it would seem that when only a part of the medium is displaced, the velocity of propagation of waves can only be increased by the velocity of the centre of gravity of the system.

This principle is evident in the case where the moving part represents exactly half of the medium; for, relating the movement of the system to its centre of gravity, which is considered for a moment as fixed, its two halves are travelling away from one another at an equal velocity in opposite directions; it follows that the waves must be slowed down in one direction as much as they are accelerated in the other, and that in relation to the centre of gravity they thus travel only at their normal velocity of propagation; or, which amounts to the same thing, they share its movement. If the moving portion were one quarter, one eighth or one sixteenth, etc., of the medium, it could be just as easily shown that the velocity to be added to the velocity of wave propagation is one quarter, one eighth, one sixteenth, and so on, of that of the part in motion—that is to say, the exact velocity of the centre of gravity; and it is clear that a theorem which holds good in all these individual instances must be generally valid.

This being established, and the prismatic medium being in equilibrium of forces *(tension)* with the surrounding ether (I am supposing for the sake of simplicity that the experiment is conducted in vacuum), any delay the light undergoes when passing through the prism when it is stationary may be considered as a result solely of its greater density, which provides a means of determining the relative densities of the two media; for we know that their relationship must be the inverse of the squares of velocity of wave propagation. Let d and d' be the wavelengths of light in the surrounding ether and in the prism, $\triangle$ and $\triangle'$ the densities of these two media; this gives us the ratio:

$$d^2 : d'^2 :: \triangle' : \triangle;$$

whence we obtain:

$$\triangle' = \triangle \frac{d^2}{d'^2},$$

and consequently:

$$\triangle' - \triangle = \triangle \left(\frac{d^2 - d'^2}{d'^2} \right).$$

This is the density of the mobile part of the prismatic medium. If the distance covered by the earth in the period of one light wave cycle is represented by t, the displacement of the centre of gravity of this medium during the same interval of time, which I am taking as a unit, or the velocity of this centre of gravity, will be:

$$t \left(\frac{d^2 - d'^2}{d^2} \right).$$

Consequently the wavelength d'' within the prism being carried along by the earth will be equal to:

$$d' + t\left(\frac{d^2 - d'^2}{d^2}\right).$$

By calculating, with the help of this expression, the interval AD' (Fig. 1.1) travelled by the ray AD before it leaves the prism, one may easily determine the direction of the refracted ray BC'. If this direction is compared to the direction of the ray BC obtained when the prism is stationary, the value for the sine of angle CBC', omitting all terms multiplied by the square or by greater powers of t because of the very small value of t, can be expressed as:

$$\frac{t}{d'} \sin i \cos i - \frac{t}{dd'} \sin i \sqrt{(d'^2 - d^2 \sin^2 i)},$$

where i represents the angle of incidence ABD.

Supposing that, from a point H somewhere on the ray BC, a line HH' is drawn parallel to the ecliptic and equal to the distance travelled by the earth during the time taken by the light to travel from B to H'; the optic axis of the telescope with which the object is being observed being directed along BH, the light must follow the direction BH' in order to arrive at H' at the same time as the cross-hair of the instrument which is being carried along with the movement of the earth: now, the line BH' coincides exactly with the direction BC' of the ray refracted by the prism, which is being carried along with the same movement; for the expression of the value of sine HBH' is also found to be:

$$\frac{t}{d'} \sin i \cos i - \frac{t}{dd'} \sin i \sqrt{(d'^2 - d^2 \sin^2 i)}.$$

Thus the telescope must be placed in the same direction as if the prism were stationary; whence it results that the movement of our earth cannot have any perceptible effect upon the apparent refraction, even supposing that it communicates only a very small part of its velocity to the ether. A very simple calculation confirms that the same is true of reflexion. Thus this hypothesis, which gives a satisfactory explanation for aberration, does not lead to any conclusion which contradicts the observed facts.

I shall conclude this letter with an application of the same theory to the experiment proposed by Boscovich, which consists in observing the phenomenon of aberration through instruments filled with water, or with some other fluid much more refractive than air, in order to ascertain whether the direction where a star is seen to lie varies as a result of the alteration in the course of the light introduced by the liquid. I shall first point out that it is unnecessary when trying to obtain this result to introduce the added complication of aberration; it can be obtained by observing a terrestrial object just as well as by observing a star. This, as it seems to me, is the simplest and most convenient way of conducting the experiment.

Having fixed relative to the instrument, or rather, to the microscope $FBDE$... [Fig. 1.2], the target M situated on the projection of the optic axis CA, this system should be placed at right angles to the ecliptic, and when the observation has been made in one direction, the whole system should be turned the other way about, and an observation made in the opposite direction. If the movement of the earth displaced the image of point M in relation to the cross-hair of the eye-piece, the image would appear now to the right, now to the left, of the cross-hair.

With the corpuscular theory it is clear, as Wilson has already pointed out, that the movement of the earth would not affect the appearance of this phenomenon. In fact, the movement of the earth means that a ray leaving M must take, in order to pass through the centre of the objective, a direction MA' such that the distance AA' is travelled by the earth in the same interval of time taken by the light to cover the distance MA', or MA (on account of the low value of the earth relative to the velocity of

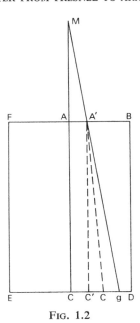

Fig. 1.2

light). When v represents the velocity of light in air, and t the velocity of the earth, we obtain:

$$MA : AA' :: v : t, \quad \text{or} \quad \frac{AA'}{MA} = \frac{t}{v};$$

which is the sine of the angle of incidence. When v' is the velocity of light in the denser medium contained in the instrument, the sine of the angle of refraction $C'A'G$ will be equal to t/v'; thus we obtain $C'G = A'C'(t/v')$; from which we derive the ratio:

$$C'G : A'C' :: t : v'.$$

Consequently the cross-hair C' of the eye-piece situated on the optic axis of the instrument will arrive at G at the same time as the light ray which has passed through the centre of the objective.

The theory of waves leads to the same results. I am supposing for the sake of simplicity that the microscope is in a vacuum. When d and d' represent the velocity of light in a vacuum and in the medium contained in the instrument, we find the value t/d for the sine of the angle of incidence AMA', and td'/d^2 for the sine of the angle of refraction $C'A'G$. Thus, independently of the displacement of the waves in the direction of the movement of the earth, $C'G = A'C'(td'/d^2)$. But the velocity with which these waves are carried along by the moving part of the medium in which they are propagated is equal to:

$$t\left(\frac{d^2-d'^2}{d^2}\right);$$

thus their total displacement Gg, during the time they take to travel through the microscope, equals:

$$\frac{A'C'}{d'}t\left(\frac{d^2-d'^2}{d^2}\right);$$

thus:

$$C'g = A'C'\cdot t\left(\frac{d'}{d^2}+\frac{d^2-d'^2}{d'd^2}\right) = A'C'\cdot t\left(\frac{d^2}{d'd^2}\right) = A'C'\cdot\frac{t}{d'}.$$

Thus we obtain the ratio $C'g : A'C' :: t : d'$; consequently the image of the point M will arrive at g at the same time as the cross-hair of the micrometer. Thus the appearance of the phenomenon must always remain the same, whatever the direction in which the instrument is turned. Although this experiment has not yet been performed, I do not doubt but that it would confirm this conclusion, which is deducible equally from the corpuscular theory and from the wave theory.

Additional note to the letter

(*Annales de chimie et de physique*, Vol. IX, p. 286, November 1818)

In calculating the refraction of light in a prism being carried along by the movement of the earth, I supposed, in order to simplify the reasoning, that the difference between the velocity of light in the prism and in the ether surrounding it was solely a result of the difference in density, elasticity being the same in the two cases; but it is very possible that the two media differ in elasticity as in density. It would even be conceivable that the elasticity of a solid body might vary according to the direction from which it was considered; and it is very probably this which gives rise to double refraction, as Dr. Young has observed. But whatever hypothesis is formed concerning the causes of the slowing of light when it passes through transparent bodies, it is always possible, in order to resolve the problem which was set me, to substitute in thought for the real medium of the prism, an elastic fluid in equilibrium of forces with the surrounding ether, and having a density such that the velocity of light is exactly the same in this fluid and in the prism, when they are supposed stationary; this equality must still remain when the two media are being carried along by the movement of the earth: these, then, are the bases upon which my calculations rest.

2. ON THE ABERRATION OF LIGHT*

G. G. STOKES

The general explanation of the phenomenon of aberration is so simple, and the coincidence of the value of the velocity of light thence deduced with that derived from the observations of the eclipses of Jupiter's satellites so remarkable, as to leave no doubt on the mind as to the truth of the explanation. But when we examine the cause of the phenomenon more closely, it is far from being so simple as it appears at first sight. On the theory of emissions, indeed, there is little difficulty; and it would seem that the more particular explanation of the cause of aberration usually given, which depends on the consideration of the motion of a telescope as light passes from its object-glass to its cross wires, has reference especially to this theory; for it does not apply to the theory of undulations, unless we make the rather startling hypothesis that the luminiferous ether passes freely through the sides of the telescope and through the earth itself. The undulatory theory of light, however, explains so simply and so beautifully the most complicated phenomena, that we are naturally led to regard aberration as a phenomenon unexplained by it, but not incompatible with it.

* Reprinted from G. G. Stokes, *Math. and Phys. Papers*, **1**, 134–40. The first part is unchanged from original (1845) publication in *Phil. Mag.*, **27**, 9; The "Additional Note" was substituted in 1880 as an unprovement of the 1845 argument to the same end.

The object of the present communication is to attempt an explanation of the cause of aberration which shall be in accordance with the theory of undulations. I shall suppose that the earth and the planets carry a portion of the ether along with them so that the ether close to their surfaces is at rest relatively to those surfaces, while its velocity alters as we recede from the surface, till, at no great distance, it is at rest in space. According to the undulatory theory, the direction in which a heavenly body is seen is normal to the fronts of the waves which have emanated from it, and have reached the neighbourhood of the observer, the ether near him being supposed to be at rest relatively to him. If the ether in space were at rest, the front of a wave of light at any instant being given, its front at any future time could be found by the method explained in Airy's tracts. If the ether were in motion, and the velocity of propagation of light were infinitely small, the wave's front would be displaced as a surface of particles of the ether. Neither of these suppositions is, however, true, for the ether moves while light is propagated through it. In the following investigation I suppose that the displacements of a wave's front in an elementary portion of time due to the two causes just considered take place independently.

Let u, v, w be the resolved parts along the rectangular axes of x, y, z, of the velocity of the particle of ether whose co-ordinates are x, y, z, and let V be the velocity of light supposing the ether at rest. In consequence of the distance of the heavenly bodies, it will be quite unnecessary to consider any waves except those which are plane, except in so far as they are distorted by the motion of the ether. Let the axis of z be taken in, or nearly in the direction of propagation of the wave considered, so that the equation of a wave's front at any time will be

$$z = C + Vt + \zeta, \qquad (1)$$

C being a constant, t the time, and ζ a small quantity, a function of x, y and t. Since u, v, w and ζ are of the order of the aberration, their squares and products may be neglected.

Denoting by α, β, γ the angles which the normal to the wave's front at the point (x, y, z) makes with the axes, we have, to the first order of approximation,

$$\cos\alpha = -\frac{d\zeta}{dx}, \quad \cos\beta = -\frac{d\zeta}{dy}, \quad \cos\gamma = 1; \qquad (2)$$

and if we take a length $V\,dt$ along this normal, the co-ordinates of its extremity will be

$$x - \frac{d\zeta}{dx} V\,dt, \quad y - \frac{d\zeta}{dy} V\,dt, \quad z + V\,dt.$$

If the ether were at rest, the locus of these extremities would be the wave's front at the time $t+dt$, but since it is in motion, the co-ordinates of those extremities must be further increased by $u\,dt$, $v\,dt$, $w\,dt$. Denoting then by x', y', z' the co-ordinates of the point of the wave's front at the time $t+dt$ which corresponds to the point (x, y, z) at the time t, we have

$$x' = x + \left(u - V\frac{d\zeta}{dx}\right) dt, \quad y' = y + \left(v - V\frac{d\zeta}{dy}\right) dt,$$
$$z' = z + (w+V)\,dt;$$

and eliminating x, y and z from these equations and (1), and denoting ζ by $f(x, y, t)$, we have for the equation to the wave's front at the time $t+dt$,

$$z' - (w+V)\,dt = C + Vt$$
$$+ f\left\{x' - \left(u + \frac{d\zeta}{dx}\right) dt, \quad y' - \left(v + \frac{d\zeta}{dy}\right) dt, \quad t\right\},$$

or, expanding, neglecting dt^2 and the square of the aberration, and suppressing the accents of x, y and z,

$$z = C + Vt + \zeta + (w+V)\,dt. \qquad (3)$$

But from the definition of ζ it follows that the equation to the wave's front at the time $t+dt$ will be got from (1) by putting $t+dt$ for t, and we have therefore for this equation

$$z = C + Vt + \zeta + \left(V + \frac{d\zeta}{dt}\right) dt. \qquad (4)$$

Comparing the identical equations (3) and (4), we have

$$\frac{d\zeta}{dt} = w.$$

This equation gives $\zeta = \int w\, dt$; but in the small term ζ we may replace $\int w\, dt$ by $\int w\, dz \div V$: this comes to taking the approximate value of z given by the equation $z = C + Vt$ instead of t for the parameter of the system of surfaces formed by the wave's front in its successive positions. Hence equation (1) becomes

$$z = C + Vt + \frac{1}{V} \int w\, dz.$$

Combining the value of ζ just found with equations (2), we get, to a first approximation,

$$\alpha - \frac{\pi}{2} = \frac{1}{V} \int \frac{dw}{dx}\, dz, \quad \beta - \frac{\pi}{2} = \frac{1}{V} \int \frac{dw}{dy}\, dz, \qquad (5)$$

equations which might very easily be proved directly in a more geometrical manner.

If random values are assigned to u, v and w, the law of aberration resulting from these equations will be a complicated one; but if u, v and w are such that $u\, dx + v\, dy + w\, dz$ is an exact differential, we have,

$$\frac{dw}{dx} = \frac{du}{dz}, \quad \frac{dw}{dy} = \frac{dv}{dz};$$

whence, denoting by the suffixes 1, 2 the values of the variables belonging to the first and second limits respectively, we obtain

$$\alpha_2 - \alpha_1 = \frac{u_2 - u_1}{V}, \quad \beta_2 - \beta_1 = \frac{v_2 - v_1}{V}. \tag{6}$$

If the motion of the ether be such that $u\,dx + v\,dy + w\,dz$ is an exact differential for one system of rectangular axes, it is easy to prove, by the transformation of co-ordinates, that it is an exact differential for any other system. Hence the formulæ (6) will hold good, not merely for light propagated in the direction first considered, but for light propagated in any direction, the direction of propagation being taken in each case for the axis of z. If we assume that $u\,dx + v\,dy + w\,dz$ is an exact differential for that part of the motion of the ether which is due to the motion of translation of the earth and planets, it does not therefore follow that the same is true for that part which depends on their motions of rotation. Moreover, the diurnal aberration is too small to the detected by observation, or at least to be measured with any accuracy, and I shall therefore neglect it.

It is not difficult to shew that the formulæ (6) lead to the known law of aberration. In applying them to the case of a star, if we begin the integrations in equations (5) at a point situated at such a distance from the earth that the motion of the ether, and consequently the resulting change in the direction of the light, is insensible, we shall have $u_1 = 0$, $v_1 = 0$; and if, moreover, we take the plane xz to pass through the direction of the earth's motion, we shall have

$$v_2 = 0, \quad \beta_2 - \beta_1 = 0,$$

and

$$\alpha_2 - \alpha_1 = \frac{u_2}{V};$$

that is, the star will appear displaced towards the direction in which the earth is moving, through an angle equal to the ratio of the velocity of the earth to that of light, multiplied by the sine of the angle between the direction of the earth's motion and the line joining the earth and the star.

Additional Note

[In what precedes *waves* of light are alone considered, and the course of a *ray* is not investigated, the investigation not being required. There follows in the original paper an investigation having for object to shew that in the case of a body like the moon or a planet which is itself in motion, the effect of the distortion of the waves in the neighbourhood of the body in altering the apparent place of the body as determined by observation is insensible. For this, the orthogonal trajectory of the wave in its successive positions from the body to the observer is considered, a trajectory which in its main part will be a straight line, from which it will not differ except in the immediate neighbourhood of the body and of the earth, where the ether is distorted by their respective motions. The perpendicular distance of the further extremity of the trajectory from the prolongation of the straight line which it forms in the intervening quiescent ether is shewn to subtend at the earth an angle which, though not actually 0, is so small that it may be disregarded.

The orthogonal trajectory of a wave in its successive positions does not however represent the course of a ray, as it would do if the ether were at rest. Some remarks made by Professor Challis in the course of discussion suggested to me the examination of the path of a ray, which in the case in which $u\,dx + v\,dy + w\,dz$ is an exact differential proved to be a straight line, a result which I had not foreseen when I wrote the above paper, which I may mention was read before the Cambridge Philosophical Society on the 18th of May, 1845 (see *Philosophical Magazine*, vol. xxix, p. 62). The rectilinearity of the path of a ray in this case, though not expressly mentioned by Professor Challis, is virtually contained in what he wrote. The problem is rather simplified by introducing the consideration of rays, and may be treated from the beginning in the following manner.

The notation in other respects being as before, let α', β' be the small angles by which the direction of the wave-normal at the point (x, y, z) deviates from that of Oz towards Ox, Oy, respectively, so that α', β' are the complements of α, β, and let $\alpha_{,}$, $\beta_{,}$ be the inclinations to Oz of the course of a ray at the same point. By compounding the velocity of propagation through the ether with the velocity of the ether we easily see that

$$\alpha_{,} = \alpha' + \frac{u}{V}, \quad \beta_{,} = \beta' + \frac{v}{V}.$$

Let us now trace the changes of $\alpha_{,}$, $\beta_{,}$ during the time dt. These depend first on the changes of α', β', and secondly on those of u, v.

As regards the change in the direction of the wave-normal, we notice that the seat of a small element of the wave in its successive positions is in a succession of planes of particles nearly parallel to the plane of x, y. Consequently the direction of the element of the wave will be altered during the time dt by the motion of the ether as much as a plane of particles of the ether parallel to the plane of the wave, or, which is the same to the order of small quantities retained, parallel to the plane xy. Now if we consider a particle of ether at the time t having for co-ordinates x, y, z, another at a distance dx parallel to the axis of x, and a third at a distance dy parallel to the axis of y, we see that the displacements of these three particles parallel to the axis of z during the time dt will be

$$w\,dt, \quad \left(w + \frac{dw}{dx}\,dx\right) dt, \quad \left(w + \frac{dw}{dy}\,dy\right) dt;$$

and dividing the relative displacements by the relation distances, we have $dw/dx \cdot dt$, $dw/dy \cdot dt$ for the small angles by which the normal is displaced, in the planes of xz, yz, from the axes x, y, so that

$$d\alpha' = -\frac{dw}{dx}\,dt, \quad d\beta' = -\frac{dw}{dy}\,dt.$$

We have seen already that the changes of u, v are $du/dz.\ V\,dt$, $dv/dz.\ V\,dt$, so that

$$d\alpha_{,} = \left(\frac{du}{dz} - \frac{dw}{dx}\right) dt, \quad d\beta_{,} = \left(\frac{dv}{dz} - \frac{dw}{dy}\right) dt.$$

Hence, provided the motion of the ether be such that

$$u\,dx + v\,dy + w\,dz$$

is an exact differential, the change of direction of a ray as it travels along is *nil*, and therefore the course of a ray is a straight line notwithstanding the motion of the ether. The rectilinearity of propagation of a ray of light, which *a priori* would seem very likely to be interfered with by the motion of the ether produced by the earth or heavenly body moving through it, is the tacit assumption made in the explanation of aberration given in treatises of Astronomy, and provided that be accounted for the rest follows as usual.[†] It follows further that the angle subtended at the earth by the perpendicular distance of the point where a ray leaves a heavenly body from the straight line prolonged which represents its course through the intervening quiescent ether, is not merely too small to be observed, but actually *nil*.]

[†] To make this explanation *quite* complete, we should properly, as Professor Challis remarks, consider the light coming from the wires of the observing telescope, in company with the light from the heavenly body.

3. ON THE RELATIVE MOTION OF THE EARTH AND THE LUMINIFEROUS ÆTHER*

A. A. MICHELSON *and* E. W. MORLEY[†]

THE discovery of the aberration of light was soon followed by an explanation according to the emission theory. The effect was attributed to a simple composition of the velocity of light with the velocity of the earth in its orbit. The difficulties in this apparently sufficient explanation were overlooked until after an explanation on the undulatory theory of light was proposed. This new explanation was at first almost as simple as the former. But it failed to account for the fact proved by experiment that the aberration was unchanged when observations were made with a telescope filled with water. For if the tangent of the angle of aberration is the ratio of the velocity of the earth to the velocity of light, then, since the latter velocity in water is three-fourths its velocity in a vacuum, the aberration observed with a water telescope should be four-thirds of its true value.[‡]

---

* *Phil. Mag.* (5) **24,** 449 (1887). This version of Michelson and Morley's article is essentially equivalent to the American version published simultaneously in *Am. J. Sci.* (3), **34,** 333 (1887).

[†] Communicated by the Authors.
This research was carried out with the aid of the Bache Fund.

[‡] It may be noticed that most writers admit the sufficiency of the explanation according to the emission theory of light; while in fact the difficulty is even greater than according to the undulatory theory. For on the emission

On the undulatory theory, according to Fresnel, first, the æther is supposed to be at rest, except in the interior of transparent media, in which, secondly, it is supposed to move with a velocity less than the velocity of the medium in the ratio $(n^2-1)/n^2$, where n is the index of refraction. These two hypotheses give a complete and satisfactory explanation of aberration. The second hypothesis, notwithstanding its seeming improbability, must be considered as fully proved, first, by the celebrated experiment of Fizeau,* and secondly, by the ample confirmation of our own work.† The experimental trial of the first hypothesis forms the subject of the present paper.

If the earth were a transparent body, it might perhaps be conceded, in view of the experiments just cited, that the intermolecular æther was at rest in space, not withstanding the motion of the earth in its orbit; but we have no right to extend the conclusion from these experiments to opaque bodies. But there can hardly be any question that the æther can and does pass through metals. Lorentz cites the illustration of a metallic barometer tube. When the tube is inclined, the æther in the space above the mercury is certainly forced out, for it is incompressible.‡ But again we have no right to assume that it makes its escape with perfect freedom, and if there be any resistance, however slight, we certainly could not assume an opaque body such as the whole earth to offer free passage through its entire mass. But as Lorentz aptly remarks:

---

theory the velocity of light must be greater in the water telescope, and therefore the angle of aberration should be less; hence, in order to reduce it to its true value, we must make the absurd hypothesis that the motion of the water in the telescope carries the ray of light in the opposite direction!

* *Comptes Rendus*, xxxiii. p. 349 (1851); *Pogg. Ann.* Ergänzungsband, iii. p. 457 (1853); *Ann. Chim. Phys.* [3], lvii. p. 385 (1859).

† "Influence of Motion of the Medium on the Velocity of Light." *Am. J. Sci.* [3], xxxi. p. 377 (1886).

‡ It may be objected that it may escape by the space between the mercury and the walls; but this could be prevented by amalgamating the latter.

"Quoi qu'il en soit, on fera bien, à mon avis, de ne pas se laisser guider, dans une question aussi importante, par des considérations sur le degré de probabilité ou de simplicité de l'une ou de l'autre hypothèse, mais de s'addresser a l'expérience pour appendre à connaître l'état, de repos ou de mouvement, dans lequel se trouve l'éther à la surface terrestre."*

In April, 1881, a method was proposed and carried out for testing the question experimentally.†

In deducing the formula for the quantity to be measured, the effect of the motion of the earth through the æther on the path of the ray at right angles to this motion was overlooked‡. The discussion of this oversight and of the entire experiment forms the subject of a very searching analysis by H. A. Lorentz§, who finds that this effect can by no means be disregarded. In consequence, the quantity to be measured had in fact but half the value supposed, and as it was already barely beyond the limits of errors of experiment, the conclusion drawn from the result of the experiment might well be questioned; since, however, the main portion of the theory remains unquestioned, it was decided to repeat the experiment with such modifications as would insure a theoretical result much too large to be masked by experimental errors. The theory of the method may be briefly stated as follows:—

Let sa, fig. 3.1, be a ray of light which is partly reflected in ab, and partly transmitted in ac, being returned by the mirrors b and c along ba and ca. ba is partly transmitted along ad, and ca is partly reflected along ad. If then the paths ab and ac are equal, the two rays interfere along ad. Suppose now, the æther being at rest, that the whole apparatus moves in the direction sc, with the velocity

* *Archives Néerlandaises*, xxi. 2me livr. *Phil. Mag.* [5], xiii. p. 236.

† "The Relative Motion of the Earth and the Luminiferous Æther," by Albert A. Michelson. *Am. J. Sci.* [3], xxii. p. 120.

‡ It may be mentioned here that the error was pointed out to the author of the former paper by M. A. Potier, of Paris, in the winter of 1881.

§ "De l'Influence du Mouvement de la Terre sur les Phen. Lum." *Archives Néerlandaises*, xxi. 2me livr. (1886).

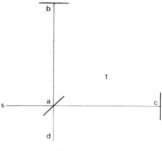

Fig. 3.1

of the earth in its orbit, the directions and distances traversed by the rays will be altered thus:—The ray *sa* is reflected along *ab*, fig. 3.2; the angle bab_i being equal to the aberration $= \alpha$, is returned along ba_i, ($aba_i = 2\alpha$), and goes to the focus of the telescope, whose direction is unaltered. The transmitted ray goes along *ac*, is returned along ca_i, and is reflected at a_i, making $ca_i e$ equal $90 - \alpha$, and therefore still coinciding with the first ray. It may be

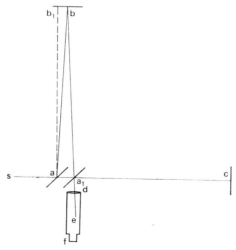

Fig. 3.2

remarked that the rays $ba_,$ and $ca_,$ do not now meet exactly in the same point $a_,$, though the difference is of the second order; this does not affect the validity of the reasoning. Let it now be required to find the difference in the two paths $aba_,$ and $aca_,$.

Let V = velocity of light.
    v = velocity of the earth in its orbit.
    D = distance ab or ac, Fig. 3.1.
    T = time light occupies to pass from a to c.
    $T_,$ = time light occupies to return from c to $a_,$ (Fig. 3.2).

Then
$$T = \frac{D}{V-v} \quad T_, = \frac{D}{V+v}.$$

The whole time of going and coming is
$$T+T_, = 2D\frac{V}{V^2-v^2},$$

and the distance travelled in this time is
$$2D\frac{V^2}{V^2-v^2} = 2D\left(1+\frac{v^2}{V^2}\right),$$

neglecting terms of the fourth order. The length of the other path is evidently
$$2D\sqrt{\left(1+\frac{v^2}{V^2}\right)},$$

or to the same degree of accuracy,
$$2D\left(1+\frac{v^2}{2V^2}\right).$$

The difference is therefore $D(v^2/V^2)$. If now the whole apparatus be turned through 90°, the difference will be in the opposite direction, hence the displacement of the interference-fringes should be $2D(v^2/V^2)$. Considering only the velocity of the earth in its orbit, this would be $2D \times 10^{-8}$. If, as was the case in the first experiment,

$D = 2 \times 10^6$ waves of yellow light, the displacement to be expected would be 0·04 of the distance between the interference-fringes.

In the first experiment, one of the principal difficulties encountered was that of revolving the apparatus without producing distortion; and another was its extreme sensitiveness to vibration. This was so great that it was impossible to see the interference-fringes except at brief intervals when working in the city, even at two o'clock in the morning. Finally, as remarked before, the quantity to be observed, namely, a displacement of something less than a twentieth of the distance between the interference-fringes, may have been too small to be detected when masked by experimental errors.

The first-named difficulties were entirely overcome by mounting the apparatus on a massive stone floating on mercury; and the second by increasing, by repeated reflexion, the path of the light to about ten times its former value.

The apparatus is represented in perspective in fig. 3.3, in plan in fig. 3.4, and in vertical section in fig. 3.5. The stone a (fig. 3.5) is about 1·5 metre square and 0·3 metre thick. It rests on an annular wooden float bb, 1·5 metre outside diameter, 0·7 metre inside diameter, and 0·25 metre thick. The float rests on mercury contained in the cast-iron trough cc, 1·5 centimetre thick, and of such dimensions as to leave a clearance of about one centimetre around the float. A pin d, guided by arms $g\,g\,g\,g$, fits into a socket e attached to the float. The pin may be pushed into the socket or be withdrawn, by a lever pivoted at f. This pin keeps the float concentric with the trough, but does not bear any part of the weight of the stone. The annular iron trough rests on a bed of cement on a low brick pier built in the form of a hollow octagon.

At each corner of the stone were placed four mirrors $dd\ ee$, fig. 3.4. Near the centre of the stone was a plane parallel glass b. These were so disposed that light from an argand burner a, passing through a lens, fell on b so as to be in part reflected to $d_{,}$; the two pencils followed the paths indicated in the figure, $b\ d\ e\ d\ b\ f$ and

$b\ d$, e, d, $b\ f$ respectively, and were observed by the telescope f. Both f and a revolved with the stone. The mirrors were of speculum metal carefully worked to optically plane surfaces five centimetres in diameter, and the glasses b and c were plane parallel of the same

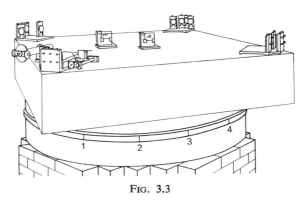

Fig. 3.3

thickness, 1·25 centimetre; their surfaces measured 5·0 by 7·5 centimetres. The second of these was placed in the path of one of the pencils to compensate for the passage of the other through the same thickness of glass. The whole of the optical portion of the apparatus was kept covered with a wooden cover to prevent air-currents and rapid changes of temperature.

The adjustment was effected as follows:—The mirrors having been adjusted by screws in the castings which held the mirrors, against which they were pressed by springs, till light from both pencils could be seen in the telescope, the lengths of the two paths measured by a light wooden rod reaching diagonally from mirror to mirror, the distance being read from a small steel scale to tenths of millimetres. The difference in the lengths of the two paths was then annulled by moving the mirror e_i. This mirror had three adjustments: it had an adjustment in altitude and one in azimuth, like all the other mirrors, but finer; it also had an adjustment in the direction of the incident ray, sliding forward or backward, but keeping very accurately parallel to its former plane. The three adjustments of this mirror could be made with the wooden cover in position.

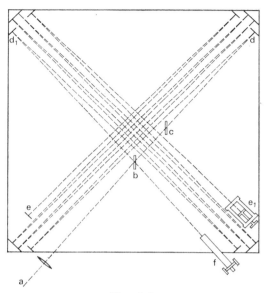

Fig. 3.4

The paths being now approximately equal, the two images of the source of light or of some well-defined object placed in front of the condensing lens, were made to coincide, the telescope was now adjusted for distinct vision of the expected interference-bands, and sodium light was substituted for white light, when the interference-bands appeared. These were now made as clear as possible by adjusting the mirror e_1; then white light was restored, the screw altering the length of path was very slowly moved (one turn of a screw of one hundred threads to the inch altering the path nearly 1000 wave-lengths) till the coloured interference-fringes reappeared in white light. These were now given a convenient width and position, and the apparatus was ready for observation.

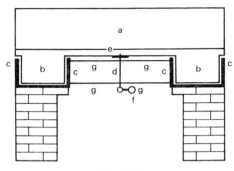

Fig. 3.5

The observations were conducted as follows:—Around the cast-iron trough were sixteen equidistant marks. The apparatus was revolved very slowly (one turn in six minutes) and after a few minutes the cross wire of the micrometer was set on the clearest of the interference-fringes at the instant of passing one of the marks. The motion was so slow that this could be done readily and accurately. The reading of the screw-head on the micrometer was noted, and a very slight and gradual impulse was given to keep up the motion of the stone; on passing the second mark, the same process was repeated, and this was continued till the apparatus had completed six revolutions. It was found that by keeping the apparatus in slow uniform motion, the results were much more uniform and consistent than when the stone was brought to rest for every observation; for the effects of strains could be noted for at least half a minute after the stone came to rest, and during this time effects of change of temperature came into action.

The following tables give the means of the six readings; the first, for observations made near noon, the second, those near six o'clock in the evening. The readings are divisions of the screw-heads.

The width of the fringes varied from 40 to 60 divisions, the mean value being near 50, so that one division means 0·02 wave-length. The rotation in the observations at noon was contrary to, and in the evening observations, in the same direction as, that of the hands of a watch.

NOON OBSERVATIONS

	16.	1.	2.	3.	4.	5.	6.	7.	8.
July 8	44·7	44·0	43·5	39·7	35·2	34·7	34·3	32·5	28·2
July 9	57·4	57·3	58·2	59·2	58·7	60·2	60·8	62·0	61·5
July 11	27·3	23·5	22·0	19·3	19·2	19·3	18·7	18·8	16·2
Mean	43·1	41·6	41·2	39·4	37·7	38·1	37·9	37·8	35·3
Mean in w.l.	·862	·832	·824	·788	·754	·762	·758	·756	·700
	·706	·692	·686	·688	·688	·678	·672	·628	·616
Final mean	·784	·762	·755	·738	·721	·720	·715	·692	·661

	9.	10.	11.	12.	13.	14.	15.	16.
July 8	26·2	23·8	23·2	20·3	18·7	17·5	16·8	13·7
July 9	63·3	65·8	67·3	69·7	70·7	73·0	70·2	72·2
July 11	14·3	13·3	12·8	13·3	12·3	10·2	7·3	6·5
Mean	34·6	34·3	34·4	34·4	33·9	33·6	31·4	30·8
Mean in w.l.	·692	·686	·688	·688	·678	·672	·628	·616
Final mean	...	...	...	...	...	...	...	...

The results of the observations are expressed graphically in fig. 3.6. The upper is the curve for the observations at noon, and the lower that for the evening observations. The dotted curves represent *one eighth* of the theoretical displacements. It seems fair to conclude from the figure that if there is any displacement due to the relative motion of the earth and the luminiferous æther, this cannot be much greater than 0·01 of the distance between the fringes.

P.M. Observations

	16.	1.	2.	3.	4.	5.	6.	7.	8.
July 8	61·2	63·3	63·3	68·2	67·7	69·3	70·3	69·8	69·0
July 9	26·0	26·0	28·2	29·2	31·5	32·0	31·3	31·7	33·0
July 12	66·8	66·5	66·0	64·3	62·2	61·0	61·3	58·7	58·4
Mean	51·3	51·9	52·5	53·9	53·9	54·1	54·3	53·7	53·4
Mean in w.l.	1·026	1·038	1·050	1·078	1·076	1·082	1·086	1·074	1·068
	1·068	1·086	1·076	1·084	1·100	1·136	1·144	1·154	1·172
Final mean	1·047	1·062	1·063	1·081	1·088	1·109	1·115	1·114	1·120

	9.	10.	11.	12.	13.	14.	15.	16.
July 8	71·3	71·3	70·5	71·2	71·2	70·5	72·5	75·7
July 9	35·8	36·5	37·3	38·8	41·0	42·7	43·7	44·0
July 12	55·7	53·7	54·7	55·0	58·2	58·5	57·0	56·0
Mean	54·3	53·8	54·2	55·0	56·8	57·2	57·7	58·6
Mean in w.l.	1·086	1·076	1·084	1·100	1·136	1·144	1·154	1·172
Final mean	...	...	...	...	...	...	...	...

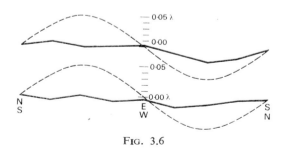

Fig. 3.6

Considering the motion of the earth in its orbit only, this displacement should be

$$2D \frac{v^2}{V^2} = 2D \times 10^{-8}.$$

The distance D was about eleven metres, or 2×10^7 wavelengths of yellow light; hence the displacement to be expected was 0·4 fringe. The actual displacement was certainly less than the twentieth part of this, and probably less than the fortieth part. But since the displacement is proportional to the square of the velocity, the relative velocity of the earth and the æther is probably less than one sixth the earth's orbital velocity, and certainly less than one fourth.

In what precedes, only the orbital motion of the earth is considered. If this is combined with the motion of the solar system, concerning which but little is known with certainty, the result would have to be modified; and it is just possible that the resultant velocity at the time of the observations was small, though the chances are much against it. The experiment will therefore be repeated at intervals of three months, and thus all uncertainty will be avoided.

It appears from all that precedes reasonably certain that if there be any relative motion between the earth and the luminiferous æther, it must be small; quite small enough entirely to refute Fresnel's explanation of aberration. Stokes has given a theory of aberration which assumes the aether at the earth's surface to be at rest with regard to the latter, and only requires in addition that the relative velocity have a potential; but Lorentz shows that these conditions are incompatible. Lorentz then proposes a modification which combines some ideas of Stokes and Fresnel, and assumes the existence of a potential, together with Fresnel's coefficient. If now it were legitimate to conclude from the present work that the æther is at rest with regard to the earth's surface, according to Lorentz there could not be a velocity potential, and his own theory also fails.

Supplement

It is obvious from what has gone before that it would be hopeless to attempt to solve the question of the motion of the solar system by observations of optical phenomena *at the surface of the earth.* But it is not impossible that at even moderate distances above the level of the sea, at the top of an isolated mountain-peak, for instance, the relative motion might be perceptible in an apparatus like that used in these experiments. Perhaps if the experiment should ever be tried under these circumstances, the cover should be of glass, or should be removed.

It may be worth while to notice another method for multiplying the square of the aberration sufficiently to bring it within the range of observation which has presented itself during the preparation of this paper. This is founded on the fact that reflexion from surfaces in motion varies from the ordinary laws of reflexion.

Let ab (fig. 1, p. 158) be a plane wave falling on the mirror mn at an incidence of 45°. If the mirror is at rest, the wave-front after reflexion will be ac.

Now suppose the mirror to move in a direction which makes an angle α with its normal, with a velocity ω. Let V be the velocity of light in the æther, supposed stationary, and let cd be the increase in the distance the light has to travel to reach d. In this time the mirror will have moved a distance

$$\frac{cd}{\sqrt{(2\cos\alpha)}}.$$

We have

$$\frac{cd}{ad} = \frac{\omega\sqrt{(2\cos\alpha)}}{V},$$

which put $= r$, and

$$\frac{ac}{ad} = 1 - r.$$

In order to find the new wave-front, draw the arc *fg* with *b* as a centre and *ad* as radius; the tangent to this arc from *d* will be the new wave-front, and the normal to the tangent from *b* will be the new direction. This will differ from the direction *ba* by θ, which it is required to find. From the equality of the triangles *adb* and *edb* it follows that $\theta = 2\phi$, $ab = ac$,

$$\tan adb = \tan\left(45° - \frac{\theta}{2}\right) = \frac{1 - \tan\dfrac{\theta}{2}}{1 + \tan\dfrac{\theta}{2}} = \frac{ac}{ad} = 1 - r,$$

or, neglecting terms of the order r^3,

$$\theta = r + \frac{r^2}{2} = \frac{\sqrt{(2\omega\cos\alpha)}}{V} + \frac{\omega^2}{V^2}\cos^2\alpha.$$

Now let the light fall on a parallel mirror facing the first, we should then have

$$\theta_{,} = \frac{-\sqrt{(2\omega\cos\alpha)}}{V} + \frac{\omega^2}{V^2}\cos_1\alpha,$$

and the total deviation would be

$$\theta + \theta_{,} = 2\varrho^2\cos^2\alpha,$$

where ϱ is the angle of aberration, if only the orbital motion of the earth is considered. The maximum displacement obtained by revolving the whole apparatus through 90° would be

$$\triangle = 2\varrho^2 = 0·004''.$$

With fifty such couples the displacement would be 0·2″. But astronomical observations in circumstances far less favourable than those in which these may be taken have been made to hundredths of a second; so that this new method bids fair to be at least as sensitive as the former.

158   NINETEENTH-CENTURY AETHER THEORIES

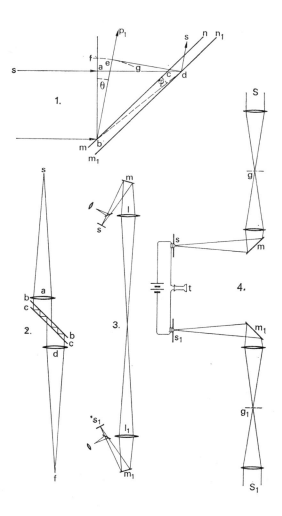

The arrangement of apparatus might be as in fig. 2; s, in the focus of the lens a, is a slit. bb, cc, are two glass mirrors optically plane, and so silvered as to allow say one twentieth of the light to pass through, and reflecting say ninety per cent. The intensity of the light falling on the observing telescope df would be about one millionth of the original intensity, so that if sunlight or the electric arc were used it could still be readily seen. The mirrors bb, and cc, would differ from parallelism sufficiently to separate the successive images. Finally, the apparatus need not be mounted so as to revolve, as the earth's rotation would be sufficient.

If it were possible to measure with sufficient accuracy the velocity of light without returning the ray to its starting point, the problem of measuring the first power of the relative velocity of the earth with respect to the æther would be solved. This may not be as hopeless as might appear at first sight, since the difficulties are entirely mechanical and may possibly be surmounted in the course of time.

For example, suppose m and $m_{\prime}$ (fig. 3) two mirrors revolving with equal velocity in opposite directions. It is evident that light from s will form a stationary image at $s_{\prime}$ and similarly light from $s_{\prime}$ will form a stationary image at s. If now the velocity of the mirrors be increased sufficiently, their phases still being exactly the same, both images will be deflected from s and $s_{\prime}$ in inverse proportion to the velocities of light in the two directions; or, if the two deflections are made equal, and the difference of phase of the mirrors be simultaneously measured, this will evidently be proportional to the difference of velocity in the two directions. The only real difficulty lies in this measurement. The following is perhaps a possible solution.

$gg_{\prime}$ (fig. 4) are two gratings on which sunlight is concentrated. These are placed so that after falling on the revolving mirrors m and $m_{\prime}$, the light forms images of the gratings at s and $s_{\prime}$, two very sensitive selenium cells in circuit with a battery and telephone. If everything be symmetrical, the sound in the telephone will be a maximum. If now one of the slits s be displaced through half the distance between the image of the grating bars, there will be silence. Suppose now that the two deflections having been made exactly equal, the slit is adjusted for silence. Then if the experiment be repeated when the earth's rotation has turned the whole apparatus through 180°, and the deflections are again made equal, there will no longer be silence, and the angular distance through which s must be moved to restore silence will measure the required difference in phase.

There remain three other methods, all astronomical, for attacking the problem of the motion of the solar system through space.

1. The telescopic observation of the proper motions of the stars. This has given us a highly probably determination of the direction of this motion, but only a guess as to its amount.

2. The spectroscopic observation of the motion of stars in the line of sight. This could furnish data for the relative motions only, though it seems likely that by the immense improvements in the photography of stellar spectra, the information thus obtained will be far more accurate than any other.

3. Finally there remains the determination of the velocity of light by observations of the eclipses of Jupiter's satellites. If the improved photometric methods practised at the Harvard observatory make it possible to observe these with sufficient accuracy, the difference in the results found for the velocity of light when Jupiter is nearest to and farthest from the line of motion will give, not merely the motion of the solar system with reference to the stars, but with reference to the luminiferous æther itself.

4. ON THE LAWS OF THE REFLEXION AND REFRACTION OF LIGHT AT THE COMMON SURFACE OF TWO NON-CRYSTALLIZED MEDIA*

G. GREEN

M. CAUCHY seems to have been the first who saw fully the utility of applying to the Theory of Light those formulæ which represent the motions of a system of molecules acting on each other by mutually attractive and repulsive forces; supposing always that in the mutual action of any two particles, the particles may be regarded as points animated by forces directed along the right line which joins them. This last supposition, if applied to those compound particles, at least, which are separable by mechanical division, seems rather restrictive; as many phenomena, those of crystallization for instance, seem to indicate certain polarities in these particles. If, however, this were not the case, we are so perfectly ignorant of the mode of action of the elements of the luminiferous ether on each other, that it would seem a safer method to take some general physical principle as the basis of our reasoning, rather than assume certain modes of action, which, after all, may be widely different from the mechanism employed by nature; more especially if this principle include in itself, as a particular case, those before used by M. Cauchy and others, and also lead to a much more simple process of calculation. The principle selected

* *Trans. Camb. Phil. Soc.* **7**, 1, 113 (1838).

as the basis of the reasoning contained in the following paper is this: In whatever way the elements of any material system may act upon each other, if all the internal forces exerted be multiplied by the elements of their respective directions, the total sum for any assigned portion of the mass will always be the exact differential of some function. But, this function being known, we can immediately apply the general method given in the *Mécanique Analytique*, and which appears to be more especially applicable to problems that relate to the motions of systems composed of an immense number of particles mutually acting upon each other. One of the advantages of this method, of great importance, is, that we are necessarily led by the mere process of the calculation, and with little care on our part, to all the equations and conditions which are *requisite* and *sufficient* for the complete solution of any problem to which it may be applied.

The present communication is confined almost entirely to the consideration of non-crystallized media; for which it is proved, that the function due to the molecular actions, in its most general form, contains only two arbitrary coefficients, A and B; the values of which depend of course on the unknown internal constitution of the medium under consideration, and it would be easy to shew, for the most general case, that any arbitrary disturbance, excited in a very small portion of the medium, would in general give rise to two spherical waves, one propagated entirely by normal, the other entirely by transverse, vibrations, and such that if the velocity of transmission of the former wave be represented by $\sqrt{A}$, that of the latter would be represented by $\sqrt{B}$. But in the transmission of light through a prism, though the wave which is propagated by normal vibrations were incapable itself of affecting the eye, yet it would be capable of giving rise to an ordinary wave of light propagated by transverse vibrations, except in the extreme cases where $A/B = 0$, or $A/B =$ a very large quantity; which, for the sake of simplicity, may be regarded as infinite; and it is not difficult to prove that the equilibrium of our medium would be unstable unless $A/B > \frac{4}{3}$. We are therefore compelled to adopt the latter value of A/B, and thus to admit that in the luminiferous ether, the velocity of transmission of waves propagated by normal vibrations is very great compared with that of ordinary light.

The principal results obtained in this paper relate to the intensity of the wave reflected at the common surface of two media, both for light polarized in and perpendicular to the plane of incidence; and likewise to the change of phase which takes place when the reflexion becomes total. In the former case, our values agree precisely with those given by Fresnel; supposing, as he has done, that the direction of the actual motion of the particles of the luminiferous ether is perpendicular to the plane of polarization. But it results from our formulæ, when the light is polarized perpendicular to the plane of incidence, that the expressions given by Fresnel are only very near approximations; and that the intensity of the reflected wave will never become absolutely null, but only attain a minimum value; which, in the case of reflexion from water at the proper angle, is $\frac{1}{151}$ part of that of the incident wave. This minimum value increases rapidly, as the index of refraction increases, and thus the quantity of light reflected at the polarizing angle, becomes considerable for highly refracting substances, a fact which has been long known to experimental philosophers.

It may be proper to observe, that M. Cauchy (*Bulletin des Sciences*, 1830) has given a method of determining the intensity of the waves reflected at the common surface of two media. He has since stated (*Nouveaux Exercises des Mathématiques*) that the hypothesis employed on that occasion is inadmissible, and has promised in a future memoir, to give a *new mechanical principle* applicable to this and other questions; but I have not been able to learn whether such a memoir has yet appeared. The first method consisted in satisfying a part, and only a part, of the conditions belonging to the surface of junction, and the consideration of the waves propagated by normal vibrations was wholly overlooked, though it is easy to perceive, that in general waves of this kind must necessarily be produced when the incident wave is polarized perpendicular to the plane of incidence, in consequence of the incident and refracted waves being in different planes. Indeed, without introducing the consideration of these last waves, it is impossible to satisfy the whole of the conditions due to the surface of junction of the two media. But when this consideration is introduced, the whole of the conditions may be satisfied, and the principles given in the *Mécanique Analytique* became abundantly sufficient for the solution of the problem.

In conclusion, it may be observed, that the radius of the sphere of sensible action of the molecular forces has been regarded as insensible with respect to the length λ of a wave of light, and thus, for the sake of simplicity, certain terms have been disregarded on which the different refrangibility of differently coloured rays might be supposed to depend. These terms, which are necessary to be considered when we are treating of the dispersion, serve only to render our formulæ uselessly complex in other investigations respecting the phenomena of light.

Let us conceive a mass composed of an immense number of molecules acting on each other by any kind of molecular forces, but which are sensible only at insensible distances, and let moreover the whole system be quite free from all extraneous action of every kind. Then x, y and z being the co-ordinates of any particle of the medium under consideration when in equilibrium, and

$$x+u, \quad y+v, \quad z+w,$$

the co-ordinates of the same particle in a state of motion (where u, v, and w are very small functions of the original co-ordinates (x, y, z), of any particle and of the time (t)), we get, by combining D'Alembert's principle with that of virtual velocities,

$$\Sigma Dm \left\{ \frac{d^2u}{dt^2} \delta u + \frac{d^2v}{dt^2} \delta v + \frac{d^2w}{dt^2} \delta w \right\} = \Sigma Dv \cdot \delta\phi \qquad (1)$$

Dm, and Dv being exceedingly small corresponding elements of the mass and volume of the medium, but which nevertheless contain a very great number of molecules, and $\delta\phi$ the exact differential of some function and entirely due to the internal actions of the particles of the medium on each other. Indeed, if $\delta\phi$ were not an exact differential, a perpetual motion would be possible, and we have every reason to think, that the forces in nature are so disposed as to render this a natural impossibility.

Let us now take any element of the medium, rectangular in a state of repose, and of which the sides are dx, dy, dz; the length of the sides composed of the same particles will in a state of motion become

$$dx' = dx(1+s_1), \quad dy' = dy(1+s_2), \quad dz' = dz(1+s_3);$$

where s_1, s_2, s_3 are exceedingly small quantities of the first order. If, moreover, we make,

$$\alpha = \cos < \frac{dy'}{dz'}, \quad \beta = \cos < \frac{dx'}{dz'}, \quad \gamma = \cos < \frac{dx'}{dy'};$$

α, β, and γ will be very small quantities of the same order. But, whatever may be the nature of the internal actions, if we represent by

$$\delta\phi \, dx \, dy \, dz,$$

the part of the second member of the equation (1), due to the molecules in the element under consideration, it is evident, that ϕ will remain the same when all the sides and all the angles of the parallelpiped, whose sides are $dx' \, dy' \, dz'$, remain unaltered, and therefore its most general value must be of the form

$$\phi = \text{function } \{s_1, s_2, s_3, \alpha, \beta, \gamma\}.$$

But s_1, s_2, s_3, α, β, γ being very small quantities of the first order, we may expand ϕ in a very convergent series of the form

$$\phi = \phi_0 + \phi_1 + \phi_2 + \phi_3 + \&\text{c}.:$$

ϕ_0, ϕ_1, ϕ_2, &c. being homogeneous functions of the six quantities α, β, γ, s_1, s_2, s_3 of the degrees 0, 1, 2, &c. each of which is very great compared with the next following one. If now, ϱ represent the primitive density of the element $dx\,dy\,dz$, we may write $\varrho\,dx\,dy\,dz$ in the place of Dm in the formula (1), which will thus become, since ϕ_2 is constant,

$$\iiint \varrho\,dx\,dy\,dz \left\{ \frac{d^2u}{dt^2}\delta u + \frac{d^2v}{dt^2}\delta v + \frac{d^2w}{dt^2}\delta w \right\}$$
$$= \iiint dx\,dy\,dz(\delta\phi_1 + \delta\phi_2 + \text{&c.});$$

the triple integrals extending over the whole volume of the medium under consideration.

But by the supposition, when $u = 0$, $v = 0$ and $w = 0$, the system is in equilibrium, and hence

$$0 = \iiint dx\,dy\,dz\,\delta\phi_1:$$

seeing that ϕ_1 is a homogeneous function of s_1, s_2, s_3, α, β, γ of the *first* degree only. If therefore we neglect ϕ_3, ϕ_4, &c. which are exceedingly small compared with ϕ_2, our equation becomes

$$\iiint \varrho\,dx\,dy\,dz \left\{ \frac{d^2u}{dt^2}\delta u + \frac{d^2v}{dt^2}\delta v + \frac{d^2w}{dt^2}\delta w \right\}$$
$$= \iiint dx\,dy\,dz\,\delta\phi_2; \qquad (2)$$

the integrals extending over the whole volume under consideration. The formula just found is true for any number of media comprised in this volume, provided the whole system be perfectly free from all extraneous forces, and subject only to its own molecular actions.

If now we can obtain the value of ϕ_2, we shall only have to apply the general methods given in the *Mécanique Analytique*.

But ϕ_2, being a homogeneous function of six quantities of the second degree, will in its most general form contain 21 arbitrary coefficients. The proper value to be assigned to each will of course depend on the internal constitution of the medium. If, however, the medium be a non-crystallized one, the form of ϕ_2 will remain the same, whatever be the directions of the co-ordinate axes in space. Applying this last consideration, we shall find that the most general form of ϕ_2 for non-crystallized bodies contains only two arbitrary coefficients. In fact, by neglecting quantities of the higher orders, it is easy to perceive that

$$s_1 = \frac{du}{dx}, \quad s_2 = \frac{dv}{dy}, \quad s_3 = \frac{dw}{dz},$$

$$\alpha = \frac{dw}{dy} + \frac{dv}{dz}, \quad \beta = \frac{dw}{dx} + \frac{du}{dz}, \quad \gamma = \frac{du}{dy} + \frac{dv}{dx},$$

and if the medium is symmetrical with regard to the plane (xy) only, ϕ_2 will remain unchanged when $-z$ and $-w$ are written for z and w. But this alteration evidently changes α and β to $-\alpha$ and $-\beta$. Similar observations apply to the planes (xz) (yz). If therefore the medium is merely symmetrical with respect to each of the three co-ordinate planes, we see that ϕ_2 must remain unaltered when

$$\left.\begin{array}{l} \text{or} \quad -z, \; -w, \; -\alpha, \; -\beta \\ \text{or} \quad -y, \; -v, \; -\alpha, \; -\gamma \\ \text{or} \quad -x, \; -u, \; -\beta, \; -\gamma \end{array}\right\} \text{ are written for } \left\{\begin{array}{l} z, \, w, \, \alpha, \, \beta \\ y, \, v, \, \alpha, \, \gamma \\ x, \, u, \, \beta, \, \gamma. \end{array}\right.$$

In this way the 21 coefficients are reduced to 9, and the resulting function is of the form

$$G\left(\frac{du}{dx}\right)^2 + H\left(\frac{dv}{dy}\right)^2 + I\left(\frac{dw}{dz}\right)^2 + L\alpha^2 + M\beta^2 + N\gamma^2$$
$$+ 2P\frac{dv}{dy} \cdot \frac{dw}{dz} + 2Q\frac{du}{dx} \cdot \frac{dw}{dz} + 2R\frac{du}{dx} \cdot \frac{dv}{dy} = \phi_2 \ldots (A).$$

Probably the function just obtained may belong to those crystals which have three axes of elasticity at right angles to each other.

Suppose now we further restrict the generality of our function by making it symmetrical all round one axis, as that of z for instance. By shifting the axis of x through the infinitely small angle $\delta\theta$,

$$\left.\begin{array}{l} x \\ y \\ z \end{array}\right\} \text{ becomes } \left\{\begin{array}{l} x + y\,\delta\theta \\ y - x\,\delta\theta, \\ z \end{array}\right.$$

$$\left.\begin{array}{l} \dfrac{d}{dx} \\ \dfrac{d}{dy} \\ \dfrac{d}{dz} \end{array}\right\} \text{ becomes } \left\{\begin{array}{l} \dfrac{d}{dx} + \delta\theta\,\dfrac{d}{dy} \\ \dfrac{d}{dy} - d\theta\,\dfrac{d}{dx}, \\ \dfrac{d}{dz} \end{array}\right.$$

and

$$\left.\begin{array}{l} u \\ v \\ w \end{array}\right\} \text{ becomes } \left\{\begin{array}{l} u + v\,\delta\theta \\ v - u\,\delta\theta. \\ w \end{array}\right.$$

Making these substitutions in (A), we see that the form of ϕ_2 will not remain the same for the new axes, unless

$$G = H = 2N + R,$$
$$L = M,$$
$$P = Q;$$

and thus we get

$$\phi_2 = G\left\{\left(\frac{du}{dx}\right)^2 + \left(\frac{dv}{dy}\right)^2\right\} + I\left(\frac{dw}{dz}\right)^2 + L(\alpha^2 + \beta^2)$$
$$+ N\gamma^2 + 2P\left(\frac{dv}{dy} + \frac{du}{dx}\right)\frac{dw}{dz} + (2G - 4N)\frac{du}{dx} \cdot \frac{dv}{dy} \ldots (B);$$

under which form it may possibly be applied to uniaxal crystals.

Lastly, if we suppose the function ϕ_2 symmetrical with respect to all three axes, there results

$$G = H = I = 2N+R,$$
$$L = M = N,$$
$$P = Q = R;$$

and consequently,

$$\phi_2 = G\left\{\left(\frac{du}{dx}\right)^2 + \left(\frac{dv}{dy}\right)^2 + \left(\frac{dw}{dz}\right)^2\right\} + L(\alpha^2+\beta^2+\gamma^2)$$
$$+ (2G-4L)\left\{\frac{dv}{dy}\cdot\frac{dw}{dz} + \frac{du}{dx}\cdot\frac{dw}{dz} + \frac{du}{dx}\cdot\frac{dv}{dy}\right\};$$

or, by merely changing the two constants and restoring the values of α, β, and γ,

$$2\phi_2 = -A\left(\frac{du}{dx} + \frac{dv}{dy} + \frac{dw}{dz}\right)^2$$
$$-B\left\{\left(\frac{du}{dy} + \frac{dv}{dx}\right)^2 + \left(\frac{du}{dz} + \frac{dw}{dx}\right)^2 + \left(\frac{dv}{dz} + \frac{dw}{dy}\right)^2\right.$$
$$\left.-4\left(\frac{dv}{dy}\cdot\frac{dw}{dz} + \frac{du}{dx}\cdot\frac{dw}{dz} + \frac{du}{dx}\cdot\frac{dv}{dy}\right)\right\}\ \ldots (C).$$

This is the most general form that ϕ_2 can take for non-crystallized bodies, in which it is perfectly indifferent in what directions the rectangular axes are placed. The same result might be obtained from the most general value of ϕ_2, by the method before used to make ϕ_2 symmetrical all round the axis of z, applied also to the other two axes. It was, indeed, thus I first obtained it. The method given in the text, however, and which is very similar to one used by M. Cauchy, is not only more simple, but has the advantage of furnishing two intermediate results, which may possibly be of use on some future occasion.

Let us now consider the particular case of two indefinitely extended media, the surface of junction when in equilibrium being a plane of infinite extent, horizontal (suppose), and which we shall

NINETEENTH-CENTURY AETHER THEORIES

take as that of (yz), and conceive the axis of x positive directed downwards. Then if ϱ be the constant density of the upper, and $\varrho_{,}$ that of the lower medium, ϕ_2 and $\phi_2^{(1)}$ the corresponding functions due to the molecular actions; the equation (2) adapted to the present case will become

$$\iiint \varrho \, dx \, dy \, dz \left\{ \frac{d^2u}{dt^2} \delta u + \frac{d^2v}{dt^2} \delta v + \frac{d^2w}{dt^2} \delta w \right\}$$
$$+ \iiint \varrho_{,} \, dx \, dy \, dz \left\{ \frac{d^2u_{,}}{dt^2} \delta u_{,} + \frac{d^2v_{,}}{dt^2} \delta v_{,} + \frac{d^2w_{,}}{dt^2} dw_{,} \right\},$$
$$= \iiint dx \, dy \, dz \, \phi_2 + \iiint dx \, dy \, dz \, \phi_2^{(1)}; \qquad (3)$$

$u_{,}$, $v_{,}$, $w_{,}$ belonging to the lower fluid, and the triple integrals being extended over the whole volume of the fluids to which they respectively belong.

It now only remains to substitute for ϕ_2 and $\phi_2^{(1)}$ their values, to effect the integrations by parts, and to equate separately to zero the coefficients of the independent variations. Substituting therefore for ϕ_2 its value (C), we get

$$\iiint dx \, dy \, dz \, \delta\phi_2$$
$$= -A \iiint dx \, dy \, dz \left\{ \left(\frac{du}{dx} + \frac{dv}{dy} + \frac{dw}{dz} \right) \left(\frac{d\delta u}{dx} + \frac{d\delta v}{dy} + \frac{d\delta w}{dz} \right) \right\}$$
$$- B \iiint dx \, dy \, dz \left\{ \left(\frac{du}{dy} + \frac{dv}{dx} \right) \left(\frac{d\delta u}{dy} + \frac{d\delta v}{dx} \right) \right.$$
$$+ \left(\frac{du}{dz} + \frac{dw}{dx} \right) \left(\frac{d\delta u}{dz} + \frac{d\delta w}{dx} \right) + \left(\frac{dv}{dz} + \frac{dw}{dy} \right) \left(\frac{d\delta v}{dz} + \frac{d\delta w}{dy} \right)$$
$$- 2 \left[\left(\frac{dv}{dy} \cdot \frac{d\delta w}{dz} + \frac{dw}{dz} \cdot \frac{d\delta v}{dy} \right) + \left(\frac{du}{dx} \cdot \frac{d\delta w}{dz} + \frac{dw}{dz} \cdot \frac{d\delta u}{dx} \right) \right.$$
$$\left. \left. + \left(\frac{du}{dx} \cdot \frac{d\delta v}{dy} + \frac{dv}{dy} \cdot \frac{d\delta u}{dx} \right) \right] \right\}$$

$$= -\iint dy\, dz \left\{ A \cdot \left(\frac{du}{dx} + \frac{dv}{dy} + \frac{dw}{dz} \right) - 2B \left(\frac{dv}{dy} + \frac{dw}{dz} \right) \right\} \cdot \delta u$$

$$- \iint dy\, dz \left\{ B \left(\frac{du}{dy} + \frac{dv}{dx} \right) \delta v + B \left(\frac{du}{dz} + \frac{dw}{dx} \right) \delta w \right\}$$

$$+ \iiint dx\, dy\, dz \left\{ A \frac{d}{dx} \cdot \left(\frac{du}{dx} + \frac{dv}{dy} + \frac{dw}{dz} \right) \right.$$

$$+ B \left[\frac{d^2u}{dy^2} + \frac{d^2u}{dz^2} - \frac{d}{dx} \left(\frac{dv}{dy} + \frac{dw}{dz} \right) \right] \right\} \cdot \delta u$$

$$+ \left\{ A \frac{d}{dy} \cdot \left(\frac{du}{dx} + \frac{dv}{dy} + \frac{dw}{dz} \right) + B \left[\frac{d^2v}{dx^2} + \frac{d^2v}{dz^2} \right. \right.$$

$$\left. \left. - \frac{d}{dy} \left(\frac{du}{dx} + \frac{dw}{dz} \right) \right] \right\} \delta v + \left\{ A \frac{d}{dz} \cdot \left(\frac{du}{dx} + \frac{dv}{dy} + \frac{dw}{dz} \right) \right.$$

$$\left. + B \left[\frac{d^2w}{dx^2} + \frac{d^2w}{dy^2} - \frac{d}{dz} \cdot \left(\frac{du}{dx} + \frac{dv}{dy} \right) \right] \right\} \delta w;$$

seeing that we may neglect the double integrals at the limits $x = -\infty$, $y = \pm \infty$, $z = \pm \infty$; as the conditions imposed at these limits cannot affect the motion of the system at any *finite* distance from the origin; and thus the double integrals belong only to the surface of junction, of which the equation, in a state of equilibrium, is

$$0 = x.$$

In like manner we get

$$\iiint dx\, dy\, dz\, \delta\phi_2^{(1)}$$

$$= + \iint dy\, dz \left\{ A_{\prime} \left(\frac{du_{\prime}}{dx} + \frac{dv_{\prime}}{dy} + \frac{dw_{\prime}}{dz} \right) - 2B_{\prime} \left(\frac{dv_{\prime}}{dy} + \frac{dw_{\prime}}{dz} \right) \right\} \delta u_{\prime}$$

$$+ \iint dy\, dz \left\{ B \left(\frac{du_{\prime}}{dy} + \frac{dv_{\prime}}{dx} \right) \delta v_{\prime} + B \left(\frac{du_{\prime}}{dz} + \frac{dw_{\prime}}{dx} \right) \delta w_{\prime} \right\}$$

+ the triple integral;

since it is the *least* value of *x* which belongs to the surface of junction in the *lower* medium, and therefore the double integrals belonging to the limiting surface must have their signs changed.

If, now, we substitute the preceding expression in (3), equate separately to zero the coefficients of the independent variation δu, δv, δw, under the triple sign of integration, there results for the upper medium

$$\varrho \frac{d^2u}{dt^2} = A \frac{d}{dx} \cdot \left(\frac{du}{dx} + \frac{dv}{dy} + \frac{dw}{dz} \right)$$
$$+ B \left\{ \frac{d^2u}{dy^2} + \frac{d^2u}{dz^2} - \frac{d}{dx} \cdot \left(\frac{dv}{dy} + \frac{dw}{dz} \right) \right\};$$

$$\varrho \frac{d^2v}{dt^2} = A \frac{d}{dy} \cdot \left(\frac{du}{dx} + \frac{dv}{dy} + \frac{dw}{dz} \right)$$
$$+ B \left\{ \frac{d^2v}{dx^2} + \frac{d^2v}{dz^2} - \frac{d}{dy} \cdot \left(\frac{du}{dx} + \frac{dw}{dz} \right) \right\}; \qquad (4)$$

$$\varrho \frac{d^2w}{dt^2} = A \frac{d}{dz} \cdot \left(\frac{du}{dx} + \frac{dv}{dy} + \frac{dw}{dz} \right)$$
$$+ B \left\{ \frac{d^2w}{dx^2} + \frac{d^2w}{dy^2} - \frac{d}{dz} \cdot \left(\frac{du}{dx} + \frac{dv}{dy} \right) \right\};$$

and by equating the coefficients of $\delta u_{,}$, $\delta v_{,}$, $\delta w_{,}$, we get three similar equations for the lower medium.

To the six general equations just obtained, we must add the conditions due to the surface of junction of the two media; and at this surface we have first,

$$u = u_{,}, \quad v = v_{,}, \quad w = w_{,}, \quad \text{(when } x = 0\text{);} \qquad (5)$$

and consequently,

$$\delta u = \delta u_{,}; \quad \delta v = \delta v_{,}; \quad \delta w = \delta w_{,}.$$

But the part of the equation (3) belonging to this surface, and which yet remains to be satisfied, is

$$0 = -\iint dy\,dz \left\{ A\left(\frac{du}{dx}+\frac{dv}{dy}+\frac{dw}{dz}\right) - 2B\left(\frac{dv}{dy}+\frac{dw}{dz}\right) \right\} \delta u$$
$$+ \iint dy\,dz \left\{ A\left(\frac{du_,}{dx}+\frac{dv_,}{dy}+\frac{dw_,}{dz}\right) - 2B_,\left(\frac{dv_,}{dy}+\frac{dw_,}{dz}\right) \right\} \delta u_,$$
$$+ \iint dy\,dz \left\{ B\left(\frac{du}{dy}+\frac{dv}{dx}\right) \delta v + B\left(\frac{du}{dz}+\frac{dw}{dx}\right) \delta w \right\}$$
$$+ \iint dy\,dz \left\{ B_,\left(\frac{du_,}{dy}+\frac{dv_,}{dx}\right) \delta v_, + B_,\left(\frac{du_,}{dz}+\frac{dw_,}{dx}\right) \delta w_, \right\};$$

and as $\delta u = \delta u_,$, &c., we obtain, as before,

$$A\left(\frac{du}{dx}+\frac{dv}{dy}+\frac{dw}{dz}\right) - 2B\left(\frac{dv}{dy}+\frac{dw}{dz}\right)$$
$$= A_,\left(\frac{du_,}{dx}+\frac{dv_,}{dy}+\frac{dw_,}{dz}\right) - 2B_,\left(\frac{dv_,}{dy}+\frac{dw_,}{dz}\right)$$
$$B\left(\frac{du}{dy}+\frac{dv}{dx}\right) = B_,\left(\frac{du_,}{dy}+\frac{dv_,}{dx}\right), \qquad (6)$$
$$B\left(\frac{du}{dz}+\frac{dw}{dx}\right) = B_,\left(\frac{du_,}{dz}+\frac{dw_,}{dx}\right);$$

and these belong to the particular value $x = 0$.

The six particular conditions (5) and (6), belonging to the surface of junction of the two media, combined with the six general equations before obtained, are *necessary* and *sufficient* for the complete determination of the motion of the two media, supposing the initial state of each given. We shall not here attempt their general solution, but merely consider the propagation of a plane wave of infinite extent, accompanied by its reflected and refracted waves, as in the preceding paper on Sound.

Let the direction of the axis of z, which yet remains arbitrary, be taken parallel to the intersection of the plane of the incident wave with the surface of junction, and suppose the disturbance of the particles to be wholly in the direction of the axis of z, which is the case with light polarized in the plane of incidence, according to Fresnel. Then we have

$$0 = u, \quad 0 = v, \quad 0 = u_{\prime}, \quad 0 = v_{\prime};$$

and supposing the disturbance the same for every point of the same front of a wave, w and $w_{\prime}$ will be independent of z, and thus the three general equations (4) will all be satisfied if

$$\varrho \frac{d^2 w}{dt^2} = B \left\{ \frac{d^2 w}{dx^2} + \frac{d^2 w}{dy^2} \right\},$$

or by making $B/\varrho = \gamma^2$,

$$\frac{d^2 w}{dt^2} = \gamma^2 \left\{ \frac{d^2 w}{dx^2} + \frac{d^2 w}{dy^2} \right\}. \tag{7}$$

Similarly in the lower medium we have

$$\frac{d^2 w_{\prime}}{dt^2} = \gamma_{\prime}^2 \left\{ \frac{d^2 w_{\prime}}{dx^2} + \frac{d^2 w_{\prime}}{dy^2} \right\}. \tag{8}$$

$w_{\prime}$ and $\gamma_{\prime}$ belonging to this medium.

It now remains to satisfy the conditions (5) and (6). But these are all satisfied by the preceding values provided

$$w = w_{\prime},$$

$$B \frac{dw}{dx} = B_{\prime} \frac{dw_{\prime}}{dx}.$$

The formulæ which we have obtained are quite general, and will apply to the ordinary elastic fluids by making $B=0$. But for all the known gases, A is independent of the nature of the gas, and consequently $A = A_{\prime}$. If, therefore, we suppose $B = B_{\prime}$, at least when we consider those phenomena only which depend merely on different states of the same medium, as is the case with light, our conditions become†

$$\left. \begin{array}{c} w = w_{,}; \\ \dfrac{dw}{dx} = \dfrac{dw_{,}}{dx} \end{array} \right\} \quad \text{(when } x = 0\text{).} \tag{9}$$

The disturbance in the upper medium which contains the incident and reflected wave, will be represented, as in the case of Sound, by

$$w = f(ax+by+ct)+F(-ax+by+ct);$$

f belonging to the incident, F to the reflected plane wave, and c being a negative quantity. Also in the lower medium,

$$w_{,} = f_{,}(a_{,}x+by+ct).$$

These values evidently satisfy the general equations (7) and (8), provided $c^2 = \gamma^2(a^2+b^2)$, and $c^2 = \gamma_{,}^2(a_{,}^2+b^2)$; we have therefore only to satisfy the conditions (9), which give

$$f(by+ct)+F(by+ct) = f_{,}(by+ct),$$

$$af'(by+ct)-aF'(by+ct) = a_{,}f_{,}'(by+ct).$$

Taking now the differential coefficient of the first equation, and writing to abridge the characteristics of the functions only, we get

$$2f' = \left(1+\frac{a_{,}}{a}\right)f_{,}', \quad \text{and} \quad 2F' = \left(1-\frac{a_{,}}{a}\right)f_{,}',$$

---

† Though for all known gases A is independent of the nature of the gas, perhaps it is extending the analogy rather too far, to assume that in the luminiferous ether the constants A and B must always be independent of the state of the ether, as found in different refracting substances. However, since this hypothesis greatly simplifies the equations due to the surface of junction of the two media, and is itself the most simple that could be selected, it seemed natural first to deduce the consequences which follow from it before trying a more complicated one, and, as far as I have yet found, these consequences are in accordance with observed facts.

and therefore

$$\frac{F'}{f'} = \frac{1-\dfrac{a_{\prime}}{a}}{1+\dfrac{a_{\prime}}{a}} = \frac{a-a_{\prime}}{a+a_{\prime}} = \frac{\cot\theta-\cot\theta_{\prime}}{\cot\theta+\cot\theta_{\prime}} = \frac{\sin(\theta_{\prime}-\theta)}{\sin(\theta_{\prime}+\theta)};$$

θ and $\theta_{\prime}$ being the angles of incidence and refraction.

This ratio between the intensity of the incident and reflected waves is exactly the same as that for light polarized in the plane of incidence (vide Airy's *Tracts*, p. 356[†]), and which Fresnel supposes to be propagated by vibrations perpendicular to the plane of incidence, agreeably to what has been assumed in the foregoing process.

We will now limit the generality of the functions f, F and $f_{\prime}$, by supposing the law of the motion to be similar to that of a cycloidal pendulum; and if we farther suppose the angle of incidence to be increased until the refracted wave ceases to be transmitted in the regular way, as in our former paper on Sound, the proper integral of the equation

$$\frac{d^2 w_{\prime}}{dt^2} = \gamma_{\prime}^2 \left\{ \frac{d^2 w_{\prime}}{dx^2} + \frac{d^2 w_{\prime}}{dy^2} \right\}$$

will be

$$w_{\prime} = \varepsilon^{-a_{\prime}'x} B \sin\psi; \tag{10}$$

where $\psi = by+ct$, and $a_{\prime}'$ is determined by

$$\gamma_{\prime}^2(b^2 - a_{\prime}'^2) = c^2 = \gamma^2(b^2 + a^2). \tag{11}$$

But one of the conditions (9) will introduce *sines* and the other *cosines*, in such a way that it will be impossible to satisfy them unless we introduce both *sines* and *cosines* into the value of w, or, which amounts to the same, unless we make

$$w = \alpha \sin(ax+by+ct+e) + \beta \sin(-ax+by+ct+e_{\prime}) \tag{12}$$

in the first medium, instead of

$$w = \alpha \sin(ax+by+ct) + \beta \sin(-ax+by+ct),$$

[†] [Airy on the Undulatory Theory of Optics, p. 109, Art. 128.]

which would have been done had the refracted wave been transmitted in the usual way, and consequently no exponential been introduced into the value of $w_{,}$. We thus see the analytical reason for what is called the change of phase which takes place when the reflexion of light becomes total.

Substituting now (10) and (12), in the equations (9), and proceeding precisely as for Sound, we get

$$0 = \alpha \cos e - \beta \cos e_{,},$$
$$0 = \alpha \sin e + \beta \sin e_{,},$$
$$\frac{a'_{,}}{a} B = \alpha \sin e - \beta \sin e_{,},$$
$$B = \alpha \cos e + \beta \cos e_{,}.$$

Hence there results $\alpha = \beta$, and $e_{,} = -e$, and

$$\tan e = \frac{a'_{,}}{a} = \frac{a'_{,}}{b} \div \frac{a}{b} = \frac{a'_{,}}{b} \tan \theta.$$

But by (11),

$$\frac{a'_{,}}{b} = \sqrt{\left\{1 - \frac{\gamma^2}{\gamma_{,}^2}\left(1 + \frac{a^2}{b^2}\right)\right\}} = \sqrt{\left(1 - \frac{1}{\mu^2 \sin^2 \theta}\right)};$$

by introducing μ the index of refraction, and θ the angle of incidence. Thus,

$$\tan e = \frac{\sqrt{(\mu^2 \sin^2 \theta - 1)}}{\mu \cos \theta};$$

and as e represents half the alteration of phase in passing from the incident to the reflected wave, we see that here also our result agrees precisely with Fresnel's for light polarized in the plane of incidence. (Vide Airy's *Tracts*, p. 362.[†])

Let us now conceive the direction of the transverse vibrations in the incident wave to be perpendicular to the direction in the case

---

† [Airy, *ubi sup.* p. 114, Art. 133.]

just considered; and therefore that the actual motions of the particles are all parallel to the intersection of the plane of incidence (xy) with the front of the wave. Then, as the planes of the incident and refracted waves do not coincide, it is easy to perceive that at the surface of junction there will, in this case, be a resolved part of the disturbance in the direction of the normal; and therefore, besides the incident wave, there will, in general, be an accompanying reflected and refracted wave, in which the vibrations are transverse, and another pair of accompanying reflected and refracted waves, in which the directions of the vibrations are normal to the fronts of the waves. In fact, unless the consideration of the two latter waves is also introduced, it is impossible to satisfy all the conditions at the surface of junction; and these are as essential to the complete solution of the problem, as the general equations of motion.

The direction of the disturbance being in plane (xy) $w=0$, and as the disturbance of every particle in the same front of a wave is the same, u and v are independent of z. Hence, the general equations (4) for the first medium become

$$\frac{d^2u}{dt^2} = g^2 \frac{d}{dx}\left(\frac{du}{dx}+\frac{dv}{dy}\right) + \gamma^2 \frac{d}{dy}\left(\frac{du}{dy}-\frac{dv}{dx}\right),$$

$$\frac{d^2v}{dt^2} = g^2 \frac{d}{dy}\left(\frac{du}{dx}+\frac{dv}{dy}\right) + \gamma^2 \frac{d}{dx}\left(\frac{dv}{dx}-\frac{du}{dy}\right),$$

where $g^2 = \dfrac{A}{\varrho}$, and $\gamma^2 = \dfrac{B}{\varrho}$.

These equations might be immediately employed in their present form; but they will take a rather more simple form, by making

$$\left.\begin{aligned}u &= \frac{d\phi}{dx} + \frac{d\psi}{dy} \\ v &= \frac{d\phi}{dy} - \frac{d\psi}{dx}\end{aligned}\right\}; \tag{13}$$

ϕ and ψ being two functions of x, y, and t, to be determined.

By substitution, we readily see that the two preceding equations are equivalent to the system

$$\left.\begin{aligned}\frac{d^2\phi}{dt^2} &= g^2\left(\frac{d^2\phi}{dx^2} + \frac{d^2\phi}{dy^2}\right) \\ \frac{d^2\psi}{dt^2} &= \gamma^2\left(\frac{d^2\psi}{dx^2} + \frac{d^2\psi}{dy^2}\right)\end{aligned}\right\}. \tag{14}$$

In like manner, if in the second medium we make

$$\left.\begin{aligned}u_{,} &= \frac{d\phi_{,}}{dx} + \frac{d\psi_{,}}{dy} \\ v_{,} &= \frac{d\phi_{,}}{dy} - \frac{d\psi_{,}}{dx}\end{aligned}\right\}, \tag{15}$$

we get to determine $\phi_{,}$ and $\psi_{,}$ the equations

$$\left.\begin{aligned}\frac{d^2\phi_{,}}{dt^2} &= g_{,}^2\left(\frac{d^2\phi_{,}}{dx^2} + \frac{d^2\phi_{,}}{dy^2}\right) \\ \frac{d^2\psi_{,}}{dt^2} &= \gamma_{,}^2\left(\frac{d^2\psi_{,}}{dx^2} + \frac{d^2\psi_{,}}{dy^2}\right)\end{aligned}\right\} \tag{16}$$

and as we suppose the constants A and B the same for both media, we have

$$\frac{\gamma}{\gamma_{,}} = \frac{g}{g_{,}}.$$

For the complete determination of the motion in question, it will be necessary to satisfy all the conditions due to the surface of junction of the two media. But, since $w = 0$ and $w_{,} = 0$, also, since u, v, $u_{,}$, $v_{,}$ are independent of z, the equations (5) and (6) become

$$u = u_{,}, \quad v = v_{,};$$

$$A\left(\frac{du}{dx} + \frac{dv}{dy}\right) - 2B\frac{dv}{dy} = A\left(\frac{du_{,}}{dx} + \frac{dv_{,}}{dy}\right) - 2B\frac{dv_{,}}{dy},$$

$$\frac{du}{dy} + \frac{dv}{dx} = \frac{du_{,}}{dy} + \frac{dv_{,}}{dx},$$

provided $x = 0$. But since $x = 0$ in the last equations, we many differentiate them with regard to any of the independent variables except x, and thus the two latter, in consequence of the two former, will become

$$\frac{du}{dx} = \frac{du_{,}}{dx}, \quad \frac{dv}{dx} = \frac{dv_{,}}{dx}.$$

Substituting now for u, v, &c., their values (13) and (15), in ϕ and ψ, the four resulting conditions relative to the surface of junction of the two media may be written,

$$\left. \begin{array}{l} \dfrac{d\phi}{dx} + \dfrac{d\psi}{dy} = \dfrac{d\phi_{,}}{dx} + \dfrac{d\psi_{,}}{dy} \\[4pt] \dfrac{d\phi}{dy} - \dfrac{d\psi}{dx} = \dfrac{d\phi_{,}}{dy} - \dfrac{d\psi_{,}}{dx} \\[4pt] \dfrac{d^2\phi}{dx^2} + \dfrac{d^2\psi}{dx\,dy} = \dfrac{d^2\phi_{,}}{dx^2} + \dfrac{d^2\psi_{,}}{dx\,dy} \\[4pt] \dfrac{d^2\phi}{dx\,dy} - \dfrac{d^2\psi}{dx^2} = \dfrac{d^2\phi_{,}}{dx\,dy} - \dfrac{d^2\psi_{,}}{dx^2} \end{array} \right\} \text{(when } x = 0\text{);}$$

or since we may differentiate with respect to y, the first and fourth equations give

$$\frac{d^2\psi}{dx^2} + \frac{d^2\psi}{dy^2} = \frac{d^2\psi_{,}}{dx^2} + \frac{d^2\psi_{,}}{dy^2};$$

in like manner, the second and third give

$$\frac{d^2\phi}{dx^2} + \frac{d^2\phi}{dy^2} = \frac{d^2\phi_{,}}{dx^2} + \frac{d^2\phi_{,}}{dy^2},$$

which, in consequence of the general equations (14) and (16), become

$$\frac{d^2\psi}{\gamma^2 \, dt^2} = \frac{d^2\psi_{,}}{\gamma_{,}^2 \, dt^2}, \quad \text{and} \quad \frac{d^2\phi}{g^2 \, dt^2} = \frac{d^2\phi_{,}}{g_{,}^2 \, dt^2}.$$

Hence, the equivalent of the four conditions relative to the surface of junction may be written

$$\left.\begin{array}{c} \dfrac{d\phi}{dx}+\dfrac{d\psi}{dy} = \dfrac{d\phi_{\prime}}{dx}+\dfrac{d\psi_{\prime}}{dy} \\ \dfrac{d\phi}{dy}-\dfrac{d\psi}{dx} = \dfrac{d\phi_{\prime}}{dy}-\dfrac{d\psi_{\prime}}{dx} \\ \dfrac{d^2\phi}{g^2\,dt^2} = \dfrac{d^2\phi_{\prime}}{g_{\prime}^2\,dt^2} \\ \dfrac{d^2\psi}{\gamma^2\,dt^2} = \dfrac{d^2\psi_{\prime}}{\gamma_{\prime}^2\,dt^2} \end{array}\right\} \text{ (when } x = 0\text{).} \qquad (17)$$

If we examine the expressions (13) and (15), we shall see that the disturbances due to ϕ and $\phi_{\prime}$ are normal to the front of the wave to which they belong, whilst those which are due to ψ, $\psi_{\prime}$ are transverse or wholly in the front of the wave. If the coefficients A and B did not differ greatly in magnitude, waves propagated by both kinds of vibrations must in general exist, as was before observed. In this case, we should have in the upper medium

and
$$\left.\begin{array}{c} \psi = f(ax+by+ct)+F(-ax+by+ct) \\ \phi = \chi_{\prime}(-a'x+by+ct) \end{array}\right\}; \qquad (18)$$

and for the lower one

$$\left.\begin{array}{c} \psi_{\prime} = f_{\prime}(a_{\prime}x+by+ct) \\ \phi_{\prime} = \chi_{\prime}(a'_{\prime}x+by+ct) \end{array}\right\}. \qquad (19)$$

The coefficients b and c being the same for all the functions to simplify the results, since the indeterminate coefficients $a'_{\prime}$ $a_{\prime}$ a' will allow the fronts of the waves to which they respectively belong, to take any position that the nature of the problem may require. The coefficient of x in F belonging to that reflected wave, which, like the incident one, is propagated by transverse vibrations would have been determined exactly like $a'_{\prime}$ $a_{\prime}$ a', as, however, it evidently $= -a$, it was for the sake of simplicity introduced immediately into our formulæ.

By substituting the values just given in the general equations (14) and (16), there results

$$c^2 = (a^2+b^2)\gamma^2 = (a_,^2+b^2)\gamma_,^2 = (a'^2+b^2)g^2 = (a_,'^2+b^2)g'^2,$$

we have thus the position of the fronts of the reflected and refracted waves.

It now remains to satisfy the conditions due to the surface of junction of the two media. Substituting, therefore, the values (18) and (19) in the equations (17), we get

$$f'' + F'' = \frac{\gamma^2}{\gamma_,^2} f_,'',$$

$$\chi'' = \frac{g^2}{g_,^2} \chi_,'';$$

$$-a'\chi' + b(f'+F') = a_,'\chi_,' + bf_,',$$
$$b\chi' - a(f'-F') = b\chi_,' - a_,f_,';$$

where to abridge, the characteristics only of the functions are written.

By means of the last four equations, we shall readily get the values of F'' χ'' $f_,''$ $\chi_,''$ in terms of f'', and thus obtain the intensities of the two reflected and two refracted waves, when the coefficients A and B do not differ greatly in magnitude, and the angle which the incident wave makes with the plane surface of junction is contained within certain limits. But in the introductory remarks, it was shewn that $A/B = $ a very great quantity which may be regarded as infinite, and therefore g and $g_,$ may be regarded as infinite compared with γ and $\gamma_,$. Hence, for all angles of incidence except such as are infinitely small, the waves dependent on ϕ and $\phi_,$ cease to be transmitted in the regular way. We shall therefore, as before, restrain the generality of our functions by supposing the law of the motion to be similar to that of a cycloidal pendulum, and as two of the waves cease to be transmitted in the regular way, we must suppose in the upper medium

$$\left.\begin{array}{l}\psi = \alpha \sin(ax+by+ct+e) + \beta \sin(-ax+by+ct+e_,)\\ \text{and}\quad \phi = \varepsilon^{a'x}(A \sin \psi_0 + B \cos \psi_0)\end{array}\right\} \quad (20)$$

and in the lower one

$$\left.\begin{array}{l}\psi_{,} = \alpha_{,} \sin(a_{,}x + by + ct) \\ \phi_{,} = \varepsilon^{-a'_{,}x}(A_{,} \sin \psi_0 + B_{,} \cos \psi_0)\end{array}\right\} \quad (21)$$

where to abridge $\psi_0 = by + ct$.

These substituted in the general equations (14) and (15), give

$$c^2 = \gamma^2(a^2 + b^2) = \gamma_{,}^2(a_{,}^2 + b^2) = g^2(-a'^2 + b^2) = g_{,}^2(-a'^2_{,} + b^2),$$

or, since g and $g_{,}$ are both infinite,

$$b = a' = a'_{,}.$$

It only remains to substitute the values (20), (21) in the equations (17), which belong to the surface of junction, and thus we get

$$bA \sin \psi_0 + bB \cos \psi_0 + b\alpha \cos(\psi_0 + e) + b\beta \cos(\psi_0 + e_{,})$$
$$= -bA_{,} \sin \psi_0 - bB_{,} \cos \psi_0 + b\alpha_{,} \cos \psi_0,$$
$$bA \cos \psi_0 - bB \sin \psi_0 - a\alpha \cos(\psi_0 + e) + a\beta \cos(\psi_0 + e_{,})$$
$$= bA_{,} \cos \psi_0 - bB_{,} \sin \psi_0 - a_{,}\alpha_{,} \cos \psi_0. \quad (22)$$
$$\frac{1}{g^2}(A \sin \psi_0 + B \cos \psi_0) = \frac{1}{g_{,}^2}(A_{,} \sin \psi_0 + B_{,} \cos \psi_0),$$
$$\frac{1}{\gamma^2}\{\alpha \sin(\psi_0 + e) + \beta \sin(\psi_0 + e_{,})\} = \frac{1}{\gamma_{,}^2}\alpha_{,} \sin \psi_0.$$

Expanding the two last equations, comparing separately the coefficients of $\cos \psi_0$ and $\sin \psi_0$, and observing that

$$\frac{g}{g_{,}} = \frac{\gamma}{\gamma_{,}} = \mu \text{ suppose,}$$

we get

$$\left.\begin{array}{l} A = \mu^2 A_{,} \\ B = \mu^2 B_{,} \\ \alpha \cos e + \beta \cos e_{,} = \mu^2 \alpha_{,} \\ \alpha \sin e + \beta \sin e_{,} = 0 \end{array}\right\}. \quad (23)$$

In like manner the two first equations of (22) will give

$$0 = A + A_{,} - \alpha \sin e - \beta \sin e_{,},$$
$$0 = A - A_{,} + \frac{a_{,}\alpha_{,}}{b} + \frac{a}{b}(\beta \cos e_{,} - \alpha \cos e),$$
$$0 = B + B_{,} + \alpha \cos e + \beta \cos e_{,} - \alpha_{,,}$$
$$0 = B - B_{,} + \frac{a}{b}(\beta \sin e_{,} - \alpha \sin e_{,});$$

combining these with the system (23), there results

$$\left. \begin{array}{l} 0 = A + A_{,} \\ 0 = B + B_{,} + (\mu^2 - 1)\alpha \\ 0 = A - A_{,} + \dfrac{a_{,}\alpha_{,}}{b} + \dfrac{a}{b}(\beta \cos e_{,} - \alpha \cos e) \\ 0 = B - B_{,} + \dfrac{a}{b}(\beta \sin e_{,} - \alpha \sin e) \end{array} \right\}. \qquad (24)$$

Again, the systems (23) and (24) readily give

$$\left. \begin{array}{l} \alpha \sin e = -\dfrac{1}{2} \cdot \dfrac{(\mu^2-1)^2}{\mu^2+1} \dfrac{b}{a} \alpha_{,} \\ \alpha \cos e = \dfrac{1}{2} \cdot \left(\mu^2 + \dfrac{a_{,}}{a}\right) \alpha_{,} \\ \beta \sin e_{,} = \dfrac{1}{2} \cdot \dfrac{(\mu^2-1)^2}{\mu^2+1} \dfrac{b}{a} \alpha_{,} \\ \beta \cos e_{,} = \dfrac{1}{2} \cdot \left(\mu^2 - \dfrac{a_{,}}{a}\right) \alpha_{,} \end{array} \right\}; \qquad (25)$$

and therefore

$$\frac{\beta^2}{\alpha^2} = \frac{(\mu^2+1)^2 \cdot \left(\mu^2 - \dfrac{a_{,}}{a}\right)^2 + (\mu^2-1)^4 \dfrac{b^2}{a^2}}{(\mu^2+1)^2 \cdot \left(\mu^2 + \dfrac{a_{,}}{a}\right)^2 + (\mu^2-1)^4 \dfrac{b^2}{a^2}}. \qquad (26)$$

When the refractive power in passing from the upper to the lower medium is not very great, μ does not differ much from 1. Hence, $\sin e$ and $\sin e_{,}$ are small, and $\cos e$, $\cos e_{,}$ do not differ sensibly from unity; we have, therefore, as a first approximation,

$$\frac{\beta}{\alpha} = \frac{\mu^2 - \dfrac{a_{,}}{a}}{\mu^2 + \dfrac{a_{,}}{a}} = \frac{\dfrac{\sin^2 \theta}{\sin^2 \theta_{,}} - \dfrac{\cot \theta_{,}}{\cot \theta}}{\dfrac{\sin^2 \theta}{\sin^2 \theta_{,}} + \dfrac{\cot \theta_{,}}{\cot \theta}} = \frac{\sin 2\theta - \sin 2\theta_{,}}{\sin 2\theta + \sin 2\theta_{,}} = \frac{\tan(\theta - \theta_{,})}{\tan(\theta + \theta_{,})},$$

which agrees with the formula in Airy's *Tracts*, p. 358[†], for light polarized perpendicular to the plane of reflexion. This result is only a near approximation: but the formula (26) gives the correct value of B^2/α^2, or the ratio of the intensity of the reflected to the incident light; supposing, with all optical writers, that the intensity of light is properly measured by the square of the actual velocity of the molecules of the luminiferous ether.

From the rigorous value (26), we see that the intensity of the reflected light never becomes absolutely null, but attains a minimum value nearly when

$$0 = \mu^2 - \frac{a_{,}}{a}, \quad \text{i.e., when} \quad \tan(\theta + \theta_{,}) = \infty, \tag{27}$$

which agrees with experiment, and this minimum value is, since (27) gives $b/a = \mu$,

$$\frac{\beta^2}{\alpha^2} = \frac{(\mu^2-1)^4 \dfrac{b^2}{a^2}}{4(\mu^2+1)^4 \, \mu^4 + (\mu^2-1)^4 \dfrac{b^2}{a^2}} = \frac{(\mu^2-1)^4}{4\mu^2(\mu^2+1)^2 + (\mu^2-1)^4} \cdots \tag{28}$$

If $\mu = \frac{4}{3}$, as when the two media are air and water, we get

$$\frac{\beta^2}{\alpha^2} = \frac{1}{151} \text{ nearly.}$$

[†] [Airy, *ubi sup.* p. 110.]

It is evident from the formula (28), that the magnitude of this minimum value increases very rapidly as the index of refraction increases, so that for highly refracting substances, the intensity of the light reflected at the polarizing angle becomes very sensible, agreeably to what has been long since observed by experimental philosophers. Moreover, an inspection of the equations (25) will shew, that when we gradually increase the angle of incidence so as to pass through the polarizing angle, the change which takes place in the reflected wave is not due to an alteration of the sign of the coefficient β, but to a change of phase in the wave, which for ordinary refracting substances is very nearly equal to 180°; the minimum value of β being so small as to cause the reflected wave sensibly to disappear. But in strongly refracting substances like diamond, the coefficient β remains so large that the reflected wave does not seem to vanish, and the change of phase is considerably less than 180°. These results of our theory appear to agree with the observations of Professor Airy. (*Camb. Phil. Trans.* Vol. IV. p. 418, &c.)

Lastly, if the velocity $\gamma_{,}$ of transmission of a wave in the lower exceed γ that in the upper medium, we may, by sufficiently augmenting the angle of incidence, cause the refracted wave to disappear, and the change of phase thus produced in the reflected wave may readily be found. As the calculation is extremely easy after what precedes, it seems sufficient to give the result. Let therefore, here, $\mu = \gamma_{,}/\gamma$, also e, $e_{,}$ and θ as before, then $e_{,} = -e$, and the accurate value of e is given by

$$\tan e = \mu \sqrt{(\mu^2 \tan^2 \theta - \sec^2 \theta)} - \frac{(\mu^2-1)^2 \tan \theta}{\mu^2+1}.$$

The first term of this expression agrees with the formula of page 362, Airy's *Tracts*[†], and the second will be scarcely sensible except for highly refracting substances.

[†] [Airy, *ubi sup.* p. 114, Art. 133.]

5. AN ESSAY TOWARDS A DYNAMICAL THEORY OF CRYSTALLINE REFLEXION AND REFRACTION*

J. MacCullagh

SECT. I. INTRODUCTORY OBSERVATIONS. EQUATION OF MOTION

Nearly three years ago I communicated to this Academy† the laws by which the vibrations of light appear to be governed in their reflexion and refraction at the surfaces of crystals. These laws—remarkable for their simplicity and elegance, as well as for their agreement with exact experiments—I obtained from a system of hypotheses which were opposed, in some respects, to notions previously received, and were not bound together by any known principles of mechanics, the only evidence of their truth being the truth of the results to which they led. On that occasion, however, I observed that the hypotheses were not independent of each other; and soon afterwards I proved that the laws of reflexion at the surface of a crystal are connected, in a very singular way, with the laws of double refraction, or of propagation in its interior; from which I was led to infer that "all these laws and hypotheses have a common source in other and more intimate laws which remain to

---

* *Transactions of the Royal Irish Academy*, **21** (1848). Read 9 December 1839.

† In a Paper "On the Laws of Crystalline Reflexion and Refraction."—*Transactions of the Royal Irish Academy*, Vol. xviii. p. 31. (*Supra,* p. 87.)

be discovered"; and that "the next step in physical optics would probably lead to those higher and more elementary principles by which the laws of reflexion and the laws of propagation are linked together as parts of the same system".† This step has since been made, and these anticipations have been realised. In the present Paper I propose to supply the link between the two sets of laws by means of a very simple theory, depending on certain special assumptions, and employing the usual methods of analytical dynamics.

In this theory, the two kinds of laws, being traced from a common origin, are at once connected with each other and severally explained; and it may be observed, that the explanation of each, as well as the source of their connexion, is now made known for the first time. For though the laws of crystalline propagation have attracted much attention during the period which has elapsed since they were discovered by Fresnel,‡ they have hitherto resisted every attempt that has been made to account for them by dynamical reasonings; and the laws of reflexion, when recently discovered, were apparently still more difficult to reach by such considerations. Nothing can be easier, however, than the process by which both systems of laws are now deduced from the same principles.

The assumptions on which the theory rests are these:—*First*, that the density of the luminiferous ether is a constant quantity; in which it is implied that this density is unchanged either by the motions which produce light or by the presence of material particles, so that it is the same within all bodies as in free space, and remains the same during the most intense vibrations. *Second*, that the vibrations in a plane-wave are rectilinear, and that, while the plane of the wave moves parallel to itself, the vibrations continue parallel to a fixed right line, the direction of this right line and the

---

† *Ibid*, p. 53, note. (*Supra*, p. 112.) The note here referred to was added some time after the Paper itself was read.

‡ These laws were published in his Memoir on Double Refraction—*Mémoires de l'Institut*, tom. vii. p. 45.

direction of a normal to the wave being functions of each other. This supposition holds in all known crystals, except quartz, in which the vibrations are elliptical.

Concerning the peculiar constitution of the ether we know nothing, and shall suppose nothing, except what is involved in the foregoing assumptions. But with respect to its physical condition generally, we shall admit, as is most natural, that a vast number of ethereal particles are contained in the differential element of volume; and, for the present, we shall consider the mutual action of these particles to be sensible only at distances which are insensible when compared with the length of a wave.

By putting together the assumptions we have made, it will appear that when a system of plane waves disturbs the ether, the vibrations are transversal, or parallel to the plane of the waves. For all the particles situated in a plane parallel to the waves are displaced, from their positions of rest, through equal spaces in parallel directions; and therefore if we conceive a closed surface of any form, including any volume great or small, to be described in the quiescent ether, and then all its points to partake of the motion imparted by the waves, any slice cut out of that volume, by a pair of planes parallel to the wave-plane and indefinitely near each other, can have nothing but its thickness altered by the displacements; and since the assumed preservation of density requires that the volume of the slice should not be altered, nor consequently its thickness, it follows that the displacements must be in the plane of the slice, that is to say, they must be parallel to the wave-plane. And conversely, when this condition is fulfilled, it is obvious that the entire volume, bounded by the arbitrary surface above described, will remain constant during the motion, while the surface itself will always contain within it the very same ethereal particles which it enclosed in the state of rest; and all this will be accurately true, no matter how great may be the magnitude of the displacements.

Let x, y, z be the rectangular co-ordinates of a particle before it is disturbed, and $x+\xi$, $y+\eta$, $z+\zeta$ its co-ordinates at the time t, the displacements ξ, η, ζ being functions of x, y, z and t. Let the ethereal density, which is the same in all media, be regarded as unity, so that *dxdydz* may, at any instant, represent indifferently either the element of volume or of mass. Then the equation of motion will be of the form

$$\int\int\int dx\,dy\,dz \left(\frac{d^2\xi}{dt^2}\delta\xi + \frac{d^2\eta}{dt^2}\delta\eta + \frac{d^2\zeta}{dt^2}\delta\zeta\right) = \int\int\int dx\,dy\,dz\,\delta V, \quad (1)$$

where V is some function depending on the mutual actions of the particles. The integrals are to be extended over the whole volume of the vibrating medium, or over all the media, if there be more than one.

Setting out from this equation, which is the general formula of dynamics applied to the case that we are considering, we perceive that our chief difficulty will consist in the right determination of the function V; for if that function were known, little more would be necessary, in order to arrive at all the laws which we are in search of, than to follow the rules of analytical mechanics, as they have been given by Lagrange. The determination of V will, of course, depend on the assumptions above stated respecting the nature of the ethereal vibrations; but, before we proceed further, it seems advisable to introduce certain lemmas, for the purpose of abridging this and the subsequent investigations.[†]

. . . . .

SECT. III. DETERMINATION OF THE FUNCTION ON WHICH THE MOTION DEPENDS. PRINCIPAL AXES OF A CRYSTAL

We come now to investigate the particular form which must be assigned to the function V, in order that the formula (1) may represent the motions of the ethereal medium. For this purpose con-

[†] [Section II is here omitted.]

ceive the plane of x' y' to be parallel to a system of plane waves whose vibrations are entirely transversal and parallel to the axis of y', so that $\xi'=0$, $\zeta'=0$. Imagine an elementary parallelpiped $dx'\,dy'\,dz'$, having its edges parallel to the axes of x', y', z', to be described in the ether when at rest, and then all its points to move according to the same law as the ethereal particles which compose it. The faces of the parallelpiped which are perpendicular to the edge dz' will be shifted, each in its own plane, in a direction parallel to the axis of y'; but their displacements will be unequal, and will differ by $d\eta'$, so that the edges connecting their corresponding angles will no longer be parallel to the axis of z', but will be inclined to it at an angle $\varkappa$ whose tangent is $d\eta'/dz'$.

Now the function V can only depend upon the directions of the axes of x', y', z' with respect to fixed lines in the crystal, and upon the angle $\varkappa$, which measures the change of form produced in the parallelpiped by vibration. This is the most general supposition which can be made concerning it. Since, however, by our second assumption, any one of these directions, suppose that of x', determines the other two, we may regard V as depending on the angle $\varkappa$ and on the direction of the axis of x' alone. But from the equations (F) it is manifest that the angle $\varkappa$ and the angles which the axis of x' makes with the fixed axes of x, y, z are all known when the quantities X, Y, Z are known. Consequently V is a function of X, Y, Z.

Supposing the angle $\varkappa$ to be very small, the quantities X, Y, Z will also be very small; and if V be expanded according to the powers of these quantities, we shall have

$$V = K+AX+BY+CZ+DX^2+EY^2+FZ^2$$
$$+GYZ+HXZ+IXY+ \&c.,$$

the quantities K, A, B, C, D, &c., being constant. But in the state of equilibrium the value of δV ought to be nothing, in whatever way the position of the system be varied; that is to say, when the displacements ξ, η, ζ, and consequently the quantities X, Y, Z, are supposed to vanish, the quantity

$$\delta V = A\delta X + B\delta Y + C\delta Z + 2DX\delta X + \&c.,$$

ought also to vanish independently of the variations $\delta\xi$, $\delta\eta$, $\delta\zeta$, or, which comes to the same thing, independently of δX, δY, δZ. Hence[†] we must have $A = 0$, $B = 0$, $C = 0$; and therefore, if we neglect terms of the third and higher dimensions, we may conclude that the variable part of V is a homogeneous function of the second degree, containing, in its general form, the squares and products of X, Y, Z, with six constant coefficients.

Of these coefficients, the three which multiply the products of the variables may always be made to vanish by changing the directions of the axes of x, y, z. For this is a known property of functions of the second degree, when the co-ordinates are the variables; and we have shown, in Lemma II., that the quantities X, Y, Z are transformed by the very same relations as the co-ordinates themselves. Therefore, in every crystal there exist three rectangular axes, with respect to which the function V contains only the squares of X, Y, Z; and as it will presently appear that the coefficients of the squares must all be negative, in order that the velocity of propagation may never become imaginary, we may consequently write, with reference to these axes,

$$V = -\tfrac{1}{2}(a^2X^2 + b^2Y^2 + c^2Z^2), \qquad (2)$$

omitting the constant K as having no effect upon the motion.

The axes of co-ordinates, in this position, are the *principal axes* of the crystal, and are commonly known by the name of *axes of elasticity*. Thus the existence of these axes is proved without any hypothesis respecting the arrangement of the particles of the medium. The constants a, b, c are the three principal velocities of propagation, as we shall see in the next section.

[†] See the reasoning of Lagrange in an analogous case *Mécanique Analytique,* tom. i. p. 68.

Having arrived at the value of V, we may now take it for the starting point of our theory, and dismiss the assumptions by which we were conducted to it. Supposing, therefore, in the first place, that a plane wave passes through a crystal, we shall seek the laws of its motion from equations (1) and (2), which contain everything that is necessary for the solution of the problem. The laws of propagation, as they are called, will in this way be deduced, and they will be found to agree exactly, so far as *magnitudes* are concerned, with those discovered by Fresnel; but the *direction* of the vibrations in a polarized ray will be different from that assigned by him. In the second place, we shall investigate the conditions which are fulfilled when light passes out of one medium into another, and we shall thus obtain the laws of reflexion and refraction at the surface of a crystal.

. . . . .

6. ON A GYROSTATIC ADYNAMIC CONSTITUTION FOR 'ETHER'*

W. Thomson (Lord Kelvin)

1. Consider the double assemblage of the red and blue atoms of § 69 of Art. xcvii. above.† Annul all the forces of attraction and of repulsion between the atoms. Join each red to its blue neighbour by a rigid bar, as in the little model which I submitted to the Academy in my last communication. We shall thus have, abutting on each red atom and on each blue, four bars making between them obtuse angles, each equal to $\pi - \cos^{-1} \frac{1}{3}$.

2. Let us suppose that each atom be a little sphere, instead of being a point; that each bar is provided at its extremities with spherical caps (as in § 70 of Art. xcvii.), rigidly fixed to it, and kept in contact with the surface of the spheres by proper guards, leaving the caps free to slide upon the spherical surfaces. We shall thus have realised an articulated molecular structure, which in aggregate constitutes a perfect incompressible quasi-liquid. The deformations must be infinitely small, and such deformations imply diminutions of volume, infinitely small and of the second order, or proportional to their squares, which we may neglect. It is because of this limitation that we have not a perfect incompressible liquid, without the qualification "quasi". But this limitation does not alter at all the perfection of our ether, so far as concerns its fitness to transmit luminous waves.

* §§ 1–6 translated from *Comptes Rendus*, 16 Sept., 1889. §§ 7–15 from *Proceedings Royal Society of Edinburgh*, 17 Mar., 1890.

† [Cf. Appendix to this sele ction forrelevant sections of Art. xcvii. K.F.S.]

3. Now to give to our structure the quasi-elasticity which it requires in order to produce the luminous waves, let us attach to each bar a gyrostatic pair composed of two Foucault gyroscopes, mounted according to the following instructions.

4. Instead of a simple bar, let us take a bar of which the central part, for a third of its length for example, is composed of two rings in planes perpendicular to one another. Let the centre of each ring, and a diameter of each ring, be in the line of the bar. Let the two rings be the exterior rings of gyroscopes, and let the axes of the interior rings be mounted perpendicularly to the line of the bar. Let us now place the interior rings, with their planes in those of the exterior rings, and consequently with the axes of their flywheels in the line of the bar. Let us give speeds of rotation, equal, but in opposite directions, to the two flywheels.

5. The gyrostatic pair thus constituted (that is to say, thus constructed and thus energised) has the singular property of requiring a Poinsot couple to be applied to the bar in order to hold it at rest in any position inclined to the position in which it was given. The moment of this couple, L, remains sensibly constant until the axes of the flywheels have turned through considerable angles from their original direction in the primitive line of the bar; and is given by the following formula which is easily demonstrated by the theory of the gyroscope,

$$L = \frac{(mk^2\omega)^2}{\mu} i,$$

i meaning the angle, supposed infinitely small, between the length of the bar in its deviated position and in its primitive position, m meaning the mass of one of the flywheels, mk^2 meaning its moment of inertia, ω meaning its angular velocity, μ meaning the moment of inertia about the axis of the pivots of the interior ring, of the entire mass (ring and flywheel) which they support.

6. Our jointed structure, with the bars placed between the black and white atoms,[†] carrying the gyrostatic pairs, is not now as

[†] [Actually red and blue atoms – K.F.S.]

formerly without rigidity; but it has an altogether peculiar rigidity, which is not like that of ordinary elastic solids, of which the forces of elasticity depend simply on the deformations which they suffer. On the contrary, its forces depend directly on the absolute rotations of the bars and only depend on the deformations, because these are kinematic consequences of the rotations of the bars. This relation of the quasi-elastic forces with absolute rotation, is just that which we require for the ether, and especially to explain the phenomena of electro-dynamics and magnetism.

7. The structure thus constituted, though it has some interest as showing a special kind of quasi-solid elasticity, due to rotation of matter having no other properties but rigidity and inertia, does not fulfil exactly the conditions of Art. xcix., § 14. The irrotational distortion of the substance or structure, regarded as a homogeneous assemblage of double points, involves essentially rotations of some of the connecting bars, and therefore requires a balancing forcive. For the 'ether' of Art. xciv. no forcive must be needed to produce any irrotational deformation: and any displacement whether merely rotational, or rotational and deformational, must require a constant couple in simple proportion to the rotation and round the same axis. In a communication to the Royal Society of Edinburgh of a year ago, I stated the problem of constructing a jointed model under gyrostatic domination to fulfil the condition of having no rigidity against irrotational deformations, and of resisting rotation, or rotational deformation, with quasi-elastic forcive in simple proportion to rotation. I gave a solution, illustrated by a model, for the case of points all in one plane; but I did not then see any very simple three-dimensional solution. After many unavailing efforts, I have recently found the following.

8. Take six fine straight rods and six straight tubes all of the same length, the internal diameter of the tubes exactly equal to the external diameter of the rods. Join all the twelve together with ends to one point P. Mechanically this might be done (but it would

not be worth the doing), by a ball-and-twelve-socket mechanism. The condition to be fulfilled is simply that the axes of the six rods and of the six tubes all pass through one point P. Make a vast number of such clusters of six tubes and six rods, and, to begin with, place their jointed ends so as to constitute an equilateral homogeneous assemblage of points $P, P', \ldots$ each connected to its twelve nearest neighbours by a rod of one sliding into a tube of the other. This assemblage of points we shall call our primary assemblage. The mechanical connections between them do not impose any constraint: each point of the assemblage may be moved arbitrarily in any direction, while all the others are at rest. The mechanical connections exist merely for the sake of providing us with rigid lines joining the points, or more properly rigid cylindric surfaces having their axes in the joining lines. Make now a rigid frame G of three rods fixed together at right angles to one another through one point O. Place it with its three bars in contact with the three pairs of rigid sides of any tetrahedron

$$(PP', P''P'''), \quad (PP'', P'''P'), \quad (PP''', P''P'),$$

of our primary assemblage. Place similarly other similar rigid frames G, G', &c., on the edges of all the tetrahedrons congener (Art. xcvii., § 13) to the one first chosen, the points $O, O', O'O''$ &c. form a second homogeneous assemblage, related to the assemblage of $P's$ just as the reds are related to the blues in Art. xcvii., § 69.

9. The position of the frame G, that is to say its orientation and the position of its centre O (six disposables) is completely determined by the four points P, P', P'', P'''. (Thomson and Tait's *Natural Philosophy*, § 198, and *Elements*, § 168.) If its bars were allowed to break away from contact with the three pairs of edges of the tetrahedrons, we might choose as its six coordinates, the six distances of its three bars from the three pairs of edges; but we suppose it to be constrained to preserve these contacts. And now let any one of the points P, P', P'', P''' or all of them be moved

in any manner, the position of the frame G is always fully determinate. This is illustrated by a model accompanying the present communication, showing a single tetrahedron of the primary assemblage and a single G frame. The edges of the tetrahedron are of copper wires sliding into glass tubes. The wires and tubes are provided with an eye or staple respectively, through which a ring passes to hold three ends together at the corners. Two of the rings have two glass tubes and one copper wire linked on each, while the other two rings have each two copper wires and one glass tube.

10. Returning now to our multitudinous assemblage, let it be displaced by stretchings of all the edges parallel to PP' with no rotation of PP' or $P''P'''$. This constitutes a homogeneous irrotational deformation of the primary assemblage. The frames G, G', &c. experience merely translatory motions without any rotation, as we see readily by confining our attention to G and the tetrahedron PP', $P''P'''$. Consider similarly five other displacements by stretchings parallel to the five other edges of the tetrahedron. Any infinitely small homogeneous deformation of the primary assemblage (§ 8 above), may be determinately resolved into six such simple stretchings, and any infinitely small rotational deformations may be produced by the superposition of a rotation without deformation, upon the irrotational deformation. Hence an infinitely small homogeneous deformation of the primary assemblage without rotation, produces only translatory motion, no rotation of the G frames: and any infinitely small homogeneous displacement whatever of the primary assemblage, produces a rotation of each frame equal to, and round the same axis as, its own rotational component.

11. It now only remains to give irrotational stability to the G frames. This may be done by mounting gyrostats properly upon them according to the principle stated in §§ 3–5 above and Art. cii. §§ 21–26 below. Three gyrostats would suffice but twelve may be taken for symmetry and for avoidance of any resultant mo-

ment of momentum of all the rotators mounted on one frame. Instead of ordinary gyrostats with rigid flywheels we may take liquid gyrostats as described below, § 12, and so make one very small step towards abolishing the crude mechanism of flywheels and axles and oiled pivots. But I chose the liquid gyrostat at present merely because it is more easily described.

12. Imagine a hollow anchor ring, or tore, that is to say an endless circular tube of circular cross-section. Perforate it in the line of a diameter and fix into it tubes to guard the perforations as shown in the accompanying diagram. Fill it with frictionless liquid, and give the liquid irrotational circulatory motion as indicated by the arrow heads in the diagram. This arrangement constitutes the hydrokinetic substitute for our mechanical flywheel. Mount it on a stiff diametral rod passing through the perforations,

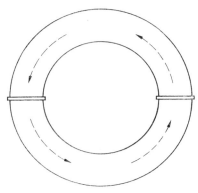

FIG. 6.1

and it becomes the mounted gyrostat, or Foucault gyroscope, required for our model. Looking back to §§ 3–4 above we see how much its use would have simplified and shortened the descriptions there given, which however was given purposely as they were because they describe real mechanism by which the exigences of our model can be practically realised in a very interesting and instructive manner, as may be seen in Art. cii., §§ 21–23 below.

13. Let XOX', YOY', ZOZ' be the three bars of the G frame: mount upon each of them four of our liquid gyrostats, those on XOX' being placed as follows and the others correspondingly. Of the four rings mounted on XX' two are to be placed in the plane of YY', XX', the other two in the plane of ZZ', XX'. The circuital fluid motions are to be in opposite directions in each pair.

14. The gyrostatic principle stated in § 5 above, applied to our G frame, with the twelve liquid gyrostats thus mounted on it, shows that if, from the position in which it was given with all the rings at rest, it be turned through an infinitesimal angle i round any axis, it requires, in order to hold it at rest in this altered position, a couple in simple proportion to i; and that this couple remains sensibly constant, as long as the planes of all the gyrostats have only changed by very small angles from parallelism to their original directions. Hence with this limitation as to time our primary homogeneous assemblage of points controlled by the gyrostatically dominated frames G, G' &c. fulfils exactly the condition stated for the ideal ether of § 14 of Art. xcix. If the velocity of the motion of the liquid in each gyrostat be infinitely great, the system exerts infinite resistance against rotation round any axis; and if the bars and tubes constituting the edges of the tetrahedron, and the bars of the G frames are all perfectly rigid, the primary assemblage is incapable of rotation or of rotational deformation: but if there is some degree of elastic flexural yielding in the edges of the tetrahedron, or in the bars of the G frame, or in all of them, the primary assemblage fulfils the definition of gyrostatic rigidity of § 14 Art. xciv. without any limit as to time, that is to say with perfect durability of its quasi-elastic rigidity.

15. A homogeneous assemblage of points with gyrostatic quasi rigidity conferred upon it in the manner described in §§ 8–14 would, if constructed on a sufficiently small scale, transmit vibrations of light exactly as does the ether of nature: and it would be incapable

of transmitting condensational-rarefactional waves, because it is absolutely devoid of resistance to condensation and rarefaction. It is in fact, a mechanical realisation of the medium to which I was led one and a half years ago,[†] from Green's original theory, by purely optical reasons, in endeavouring to explain results of observation regarding the refraction and reflection of light.

APPENDIX

Being article XCVII, §§ 66–70, of W. Thomson's Math. and Phys. Papers, iii. 425 ff.

§ 66. Try first to realise an incompressible elastic solid. When this is done we shall see, by an inevitably obvious modification, how to give any degree of compressibility we please without changing the rigidity, and so to realise an elastic solid with any given positive rigidity, and any given positive or negative bulk-modulus (stable without any surface constraint, only when the bulk-modulus is positive).

§ 67. To aid conception, make a tetrahedronal model of six equal straight rods, jointed at the angular points in which three meet, each having longitudinal elasticity with perfect anti-flexural rigidity. These constitute merely an ideal materialisation of the connection assumed in the Boscovich attractions and repulsions. A very telling *realisation* of the system thus imagined is made by taking six equal and similar bent bows and jointing their ends together by threes. The jointing might be done accurately by a ball and double socket mechanism of an obvious kind, but it would not be worth the doing. A rough arrangement of six bows of bent steel wire, merely linked together by hooking an end of one into rings on the ends of two others, may be made in a few minutes; and even its defects are not unhelpful towards a vivid understanding of our subject. We have now an element of elastic

[†] *Philosophical Magazine*, Nov. 1888, On the reflection and refraction of light, by Sir W. Thomson.

solid which clearly has an essentially definite ratio of compressibility to reciprocal of either of the rigidities (§ 27 above), each being inversely proportional to the stiffness of the bows. Now we can obviously make this solid incompressible if we take a boss jointed to four equal tie-struts, and joint their free ends to the four corners of the tetrahedron; and we do not alter either of the rigidities if the length of each tie-strut is equal to distance from centre to corners of the unstressed tetrahedron. If the tie-struts are shorter than this, their effect is clearly to augment the rigidities; if longer, to diminish the rigidities. The mathematical investigation proves that it diminishes the greater of the rigidities more than it diminishes the less, and that before it annuls the less it equalises the greater to it.

§ 68. If for the present we confine our attention to the case of the tie-struts longer than the un-strained distance from centre to corners, simple struts will serve; springs, such as bent bows, capable of giving thrust as well as pull along the sides of the tetrahedron, are not needed; mere india-rubber elastic filaments will serve instead, or ordinary spiral springs, and all the end-jointings become much simplified. A realised model accompanies this communication.

§ 69. The model being completed, we have two simple homogeneous Bravais assemblages of points; reds and blues, as we shall call them for brevity; so placed that each blue is in the centre of a tetrahedron of reds, and each red in the centre of a tetrahedron of blues. The other tetrahedronal groupings (Molecular Tactics, §§ 45, 60) being considered, each tetrahedron of reds is vacant of blue, and each tetrahedron of blues is vacant of reds.[†]

[†] An interesting structure is suggested by adding another homogeneous assemblage, marked green; giving a green in the centre of each hitherto vacant tetrahedron of reds. It is the same assemblage of triplets as that described in § 24 above. It does not (as long as we have mere jointed struts of constant length between the greens and reds) modify our rigidity-modulus, nor otherwise help us at present, so, having inevitably noticed it, we leave it.

§ 70. Imagine the springs removed and the struts left; but now all properly jointed by fours of ends with perfect frictionless ball-and-socket triple-joints. We have a perfectly non-rigid three-dimensional skeleton frame-work, analogous to idealised plane netting consisting of stiff straight sides of hexagons perfectly jointed in threes of ends. [Compare Art. C., § 2, below.]

7. ON THE ELECTROMAGNETIC THEORY OF THE REFLECTION AND REFRACTION OF LIGHT*

G. F. FITZGERALD

IN THE second volume of his "Electricity and Magnetism", Professor J. Clerk Maxwell has proposed a very remarkable electromagnetic theory of light, and has worked out the results as far as the transmission of light through uniform crystalline and magnetic media are concerned, leaving the questions of reflection and refraction untouched. These, however, may be very conveniently studied from his point of view.

If we call W the electrostatic energy of the medium, it may be expressed in terms of the electromotive force and the electric displacement at each point, as is done in Professor Maxwell's "Electricity and Magnetism", vol. ii, part 4, ch. 9. I shall adopt his notation, and call the electromotive force $\mathfrak{E}$, and its components P, Q, R; and the electric displacement $\mathfrak{D}$, and its components f, g, h. As several of the results of this Paper admit of a very elegant expression in Quaternion notation, I shall give the work and results in both Cartesian and Quaternion form, confining the German letters to the Quaternion notation. Between these quantities, then,

---

* From the *Philosophical Transactions of the Royal Society* (Pt. II, 1880; Art. xix, p. 691). Communicated by G. J. Stoney, M.A., F.R.S., Secretary of the Queen's University in Ireland. Received 26 October, 1878. Read 9 January, 1879.

we have the equation

$$W = -\tfrac{1}{2} \iiint S\mathfrak{E}\mathfrak{D}\, dx\, dy\, dz = \tfrac{1}{2} \iiint (Pf+Qg+Rh)\, dx\, dy\, dz.$$

Similarly, the kinetic energy T may be expressed in terms of the magnetic induction $\mathfrak{B}$, and the magnetic force $\mathfrak{H}$, or their components a, b, c, and a, β, γ, by the equation

$$T = -\frac{1}{8\pi} \iiint S\mathfrak{B}\mathfrak{H}\, dx\, dy\, dz$$
$$= \frac{1}{8\pi} \iiint (a\alpha+b\beta+c\gamma)\, dx\, dy\, dz.$$

I shall at present assume this to be a complete expression for T, and return to the case of magnetized media for separate treatment, as Professor Maxwell has proposed additional terms in this case, in order to account for their property of rotatory polarization. I shall throughout assume the media to be isotropic as regards magnetic induction, for the contrary supposition would enormously complicate the question, and be, besides, of doubtful physical applicability. For the present I shall not assume them to be electrostatically isotropic. Hence $\mathfrak{E}$ is a linear vector and self-conjugate function of $\mathfrak{D}$, and consequently P, Q, R linear functions of f, g, h, so that we may write in Quaternion notation

$$\mathfrak{E} = \phi\mathfrak{D};$$

and if we call U the general symmetrical quadratic function of f, g, h, we may assume

$$U = Pf+Qg+Rh,$$

and consequently

$$W = -\tfrac{1}{2} \iiint S\mathfrak{D}\phi\mathfrak{D}\, dx\, dy\, dz = \tfrac{1}{2} \iiint U\, dx\, dy\, dz.$$

As the medium is magnetically isotropic, we have

$$\mathfrak{B} = \mu\mathfrak{H}, \quad \text{or} \quad a = \mu\alpha, \quad b = \mu\beta, \quad c = \mu\gamma,$$

where μ is the coefficient of magnetic inductive capacity, and consequently the electrokinetic energy may be written

$$T = -\frac{\mu}{8\pi} \iiint \mathfrak{H}^2 \, dx \, dy \, dz = \frac{\mu}{8\pi} \iiint (\alpha^2 + \beta^2 + \gamma^2) \, dx \, dy \, dz.$$

Now I shall assume the mediums to be non-conductors; and although this limits to some extent the applicability of my results, and notably their relation to metallic reflection, yet it is a necessity, for otherwise the problem would be beyond my present powers of solution. With this assumption, and using Newton's notation of $\dot{x}$ for dx/dt, we have the following equations (see "Electricity and Magnetism", vol. ii., § 619):—

$$4\pi\dot{\mathfrak{D}} = V \nabla \mathfrak{H},$$

using ∇ for the operation

$$i\frac{d}{dx} + j\frac{d}{dy} + k\frac{d}{dz};$$

or the same in terms of its components, namely,

$$4\pi\dot{f} = \frac{d\gamma}{dy} - \frac{d\beta}{dz},$$

$$4\pi\dot{g} = \frac{d\alpha}{dz} - \frac{d\gamma}{dx},$$

$$4\pi\dot{h} = \frac{d\beta}{dx} - \frac{d\alpha}{dy}.$$

Assuming now a quantity $\mathfrak{R}$ with components ξ, η, ζ, such that

$$\mathfrak{R} = \int \mathfrak{H} \, dt,$$

and consequently

$$\dot{\mathfrak{R}} = \mathfrak{H};$$

or, in terms of the components,
$$\dot{\xi} = \alpha, \quad \dot{\eta} = \beta, \quad \dot{\zeta} = \gamma,$$
we may evidently write
$$4\pi \mathfrak{D} = \mathrm{V} \triangledown \mathfrak{R},$$
i.e., $\quad 4\pi f = \dfrac{d\zeta}{dy} - \dfrac{d\eta}{dz}, \quad 4\pi g = \dfrac{d\xi}{dz} - \dfrac{d\zeta}{dx}, \quad 4\pi h = \dfrac{d\eta}{dx} - \dfrac{d\xi}{dy},$

so that we have
$$\mathrm{W} = -\frac{1}{32\pi^2} \iiint \mathrm{S}(\mathrm{V} \triangledown \mathfrak{R} \cdot \phi \mathrm{V} \triangledown \mathfrak{R})\, dx\, dy\, dz,$$
$$\mathrm{T} = -\frac{\mu}{8\pi} \iiint \dot{\mathfrak{R}}^2\, dx\, dy\, dz = \frac{\mu}{8\pi} \iiint (\dot{\xi}^2 + \dot{\eta}^2 + \dot{\zeta}^2)\, dx\, dy\, dz.$$

Lagrange's equations of motion may often be very conveniently represented as the conditions that $\int(\mathrm{T}-\mathrm{W})\, dt$ should be a minimum, or, in other words, that

$$\delta \int (\mathrm{T} - \mathrm{W})\, dt = 0,$$

and this method, from its symmetry, is particularly applicable to the methods of Quaternions.

. . . . .

8. THE ROTATIONAL ETHER IN ITS APPLICATION TO ELECTROMAGNETISM*

O. HEAVISIDE

ACCORDING to Maxwell's theory of electric displacement, disturbances in the electric displacement and magnetic induction are propagated in a non-conducting dielectric after the manner of motions in an incompressible solid. The subject is somewhat obscured in Maxwell's treatise by his equations of propagation containing A, Ψ, J, all of which are functions considerably remote from the vectors which represent the state of the field, viz., the electric and magnetic forces, and by some dubious reasoning concerning Ψ and J. There is, however, no doubt about the statement with which I commenced, as it becomes immediately evident when we ignore the potentials and use E or H instead, the electric or the magnetic force.

The analogy has been made use of in more ways than one, and can be used in very many ways. The easiest of all is to assume that the magnetic force is the velocity of the medium, magnetic induction the momentum, and so on, as is done by Prof. Lodge (Appendix to "Modern Views of Electricity"). I have also used this method for private purposes, on account of the facility with which electromagnetic problems may be made elastic-solid problems. I have shown that when impressed electric force acts it is the curl or rotation of the electric force which is to be considered

* O. Heaviside *Electromagnetic Theory*, Dover edit: New York, 1950. Originally published in the January, 1891 issue of *The Electrician*, **26**, 360.

as the source of the resulting disturbances. Now, on the assumption that the magnetic force is the velocity in the elastic solid, we find that the curl of the impressed electric force is represented simply by impressed mechanical force of the ordinary Newtonian type. This is very convenient.

But the difficulties in the way of a complete and satisfactory representation of electromagnetic phenomena, by an elastic-solid ether are insuperable. Recognising this, Sir W. Thomson has recently brought out a new ether; a rotational ether. It is incompressible, and has no true rigidity, but possesses a quasi-rigidity arising from elastic resistance to absolute rotation.

The stress consists partly of a hydrostatic pressure (which I shall ignore later), but there is no distorting stress, and its place is taken by a rotating stress. It gives rise to a translational force and a torque. If **E** be the torque, the stress on any plane N (unit normal) is simply **VEN**, the vector product of the torque and the normal vector.

The force is $-\text{curl } \mathbf{E}$. We have therefore the equation of motion

$$-\text{curl } \mathbf{E} = \mu\dot{\mathbf{H}},$$

if H is the velocity and μ the density. But, alas, the torque is proportional to the rotation. This gives

$$\text{curl } \mathbf{H} = c\dot{\mathbf{E}},$$

where c is the compliancy, the reciprocal of the quasi-rigidity.

Now these are the equations connecting electric and magnetic force in a non-conducting dielectric, when μ is the inductivity and c the permittancy. We have a parallelism in detail, not merely in some particulars. The kinetic energy $\frac{1}{2}\mu H^2$ represents the magnetic energy, and the potential energy $\frac{1}{2}cE^2$ the electric energy. The vector-flux of energy is **VEH**, the activity of the stress.

This mode of representation differs from that of Sir W. Thomson, who represents magnetic force by rotation. This system makes electric energy kinetic, and magnetic energy potential, which I do not find so easy to follow.

Now let us, if possible, extend our analogy to conductors. Let the translational and the rotational motions be both frictionally resisted, and let the above equations become

$$-\text{curl } \mathbf{E} = g\mathbf{H} + \mu\dot{\mathbf{H}},$$
$$\text{curl } \mathbf{H} = k\mathbf{E} + c\dot{\mathbf{E}},$$

where g is the translational frictionality; k will be considered later. We have now the equations of electric and magnetic force in a dielectric with duplex conductivity, k being the electric and g the magnetic conductivity (by analogy with electric force, but a frictionality in our present dynamical analogy).

We have, therefore, still a parallelism in every detail. We have waste of energy by friction $g\mathrm{H}^2$ (translational) and $k\mathrm{E}^2$ (rotational). If $g/\mu = k/c$ the propagation of disturbances will take place precisely as in a non-conducting dielectric, though with attenuation caused by the loss of energy.

To show how this analogy works out in practice, consider a telegraph circuit, which is most simply taken to be three co-axial tubes. Let, A, B, and C be the tubes; A the innermost, C the outermost, B between them; all closely fitted. Let their material be the rotational ether. In the first place, suppose that there is perfect slip between B and its neighbours. Then, when a torque is applied to the end of B (the axis of torque to be that of the tubes), and circular motion thus given to B, the motion is (in virtue of the perfect slip) transmitted along B, without change of type, at constant speed, and without affecting A and C.

This is the analogue of a concentric cable, if the conductors A and C be perfect conductors, and the dielectric B a perfect insulator. The terminal torque corresponds to the impressed voltage. It should be so distributed over the end of B that the applied force there is circular tangential traction, varying inversely as the distance from the axis; like the distribution of magnetic force, in fact.

Now, if we introduce translational and rotational resistance in B, in the above manner, still keeping the slip perfect, we make the dielectric not only conducting electrically but also magnetically.

This will not do. Abolish the translational resistance in B altogether, and let there be no slip at all between B and A, and B and C. Let also there be rotational resistance in A and C.

We have now the analogue of a real cable: two conductors separated by a third. All are dielectrics, but the middle one should have practically very slight conductivity, so that it is pre-eminently a dielectric; whilst the other two should have very high conductivity, so that they are pre-eminently conductors. The three constants, μ, c, k, may have any value in the three tubes, but practically k should be in the middle tube a very small fraction of what it is in the others.

It is remarkable that the *quasi-rotational* resistance in A and C should tend to counteract the distorting effect on waves of the *quasi-rotational* resistance in B. But the two rotations, it should be observed, are practically perpendicular, being axial or longitudinal (now) in A and C, and transverse or radial in B; due to the relative smallness of k in the middle tube.

To make this neutralising property work exactly we must transfer the resistance in the tubes A and C to the tube B, at the same time making it translational resistance. Also restore the slip. Then we can have perfect annihilation of distortion in the propagation of disturbances, viz., when k and g are so proportioned as to make the two wastes of energy equal. In the passage of a disturbance along B there is partial absorption, but no reflection.

But as regards the meaning of the above k there is a difficulty. In the original rotational ether the torque varies as the rotation. If we superadd a real frictional resistance to rotation we get an equation of the form

$$\dot{\mathbf{E}} = \left(a + b\frac{d}{dt}\right) \operatorname{curl} \mathbf{H},$$

E being (as before) the torque, and **H** the velocity. But this is not of the right form, which is (as above)

$$\operatorname{curl} \mathbf{H} = \left(k + c\frac{d}{dt}\right)\mathbf{E};$$

therefore some special arrangement is required (to produce the dissipation of energy kE^2), which does not obviously present itself in the mechanics of the rotational ether.

On the other hand, if we follow up the other system, in which magnetic force is allied with rotation, we may put $g = 0$, let $-\mathbf{E}$ be the velocity and $\mathbf{H}$ the torque; μ the compliancy, c the density, and k the translational frictionality. This gives

$$-\text{curl } \mathbf{E} = \mu \dot{\mathbf{H}}$$

$$\text{curl } \mathbf{H} = k\mathbf{E} + c\dot{\mathbf{E}}.$$

We thus represent a homogeneous conducting dielectric, with a translational resistance to cause the Joulean waste of energy. But it is now seemingly impossible to properly satisfy the conditions of continuity at the interface of different media. For instance, the velocity $-\mathbf{E}$ should be continuous, but we do not have normal continuity of electric force at an interface. In the case of the tubes we avoided this difficulty by having the velocity tangential.

Either way, then, the matter is left, for the present, in an imperfect state.

In the general case, the d/dt of our equations should receive an extended meaning, on account of the translational motion of the medium. The analogy will, therefore, work out less satisfactorily. And it must be remembered that it is only an analogy in virtue of similitude of relations. We cannot, for instance, deduce the Maxwellian stresses and mechanical forces on charged or currented bodies. The similitude does not extend so far. But certainly the new ether goes somewhat further than anything known to me that has been yet proposed in the way of a stressed solid.

[P.S.—The special reckonings of torque and rotation in the above are merely designed to facilitate the elastic-solid and electromagnetic comparisons without unnecessary constants.]

9. AETHER AND MATTER*

J. Larmor

DYNAMICAL THEORY OF ELECTRICAL ACTIONS

Least Action, fundamental in General Dynamics

49. The idea of deducing all phenomenal changes from a principle of least expenditure of effort or action dates for modern times, as is well known, from the speculations of Maupertuis. The main illustration with which he fortified his view was Fermat's principle of least time for ray propagation in optics. This optical law follows as a direct corollary from Huygens' doctrine that radiation is propagated by wave-motions. In Maupertuis' hands, however, it reverted to the type of a dogma of least action in the dynamical sense as originally enunciated vaguely by Descartes, which Fermat's statement of the principle as one of least time was intended to supersede; under that aspect it was dynamically the equally immediate corollary of the corpuscular theory of optical rays which was finally adopted by Newton.

The general idea of Maupertuis at once attracted the attention of mathematicians; and the problem of the exact specification of the Action, so as to fulfil the minimum relation, was solved by Euler for the case of orbits of particles. Shortly afterwards the

* *Aether and Matter*, published by Cambridge University Press, Cambridge, 1900.

solution was re-stated with greater precision, and generalized to all material systems, by Lagrange (*Mem. Taurin.*, 1760) in one of his earliest and most brilliant memoirs, which constructed the algorithm of the Calculus of Variations, and at the same time also laid the foundation of the fundamental physical science of Analytical Dynamics. The subsequent extensions by Hamilton of the Lagrangian analytical procedure involve, so far as interpretation has hitherto been enabled to go, rather fundamental developments in the mathematical methods than new physical ideas,—except in the weighty result that the mere expression of all the quantities of the system as differential coefficients of a single characteristic function establishes relations of complete reciprocity between them, and also between the various stages, however far apart in time, of the system's progress.

It is now a well-tried resource to utilize the principle that every dynamical problem can be enunciated, in a single formula, as a variation problem, in order to help in the reduction to dynamics of physical theories in which the intimate dynamical machinery is more or less hidden from direct inspection. If the laws of any such department of physics can be formulated in a minimum or variational theorem, that subject is thereby virtually reduced to the dynamical type: and there remain only such interpretations, explanations, and developments, as will correlate the integral that is the subject of variation with the corresponding integrals relating to known dynamical systems. These developments will usually take the form of the tracing out of analogies between the physical system under consideration and dynamical systems which can be directly constructed to have Lagrangian functions of the same kind: they do not add anything logically to the completeness and sufficiency of the analytical specification of the system, but by being more intuitively grasped by the mind and of more familiar type, they often lead to further refinements and developments which carry on our theoretical views into still higher and more complete stages.

Derivation of the Equations of the Electric Field from the Principle of Least Action

50. It has been seen (§ 48) that the only effective method of working out the dynamics of molecular systems is to abolish the idea of force between the molecules, about which we can directly know nothing, and to formulate the problem as that of the determination of the natural sequence of changes of configuration in the system. If the individual molecules are to be permanent, the system, when treated from the molecular standpoint, must be conservative; so that the Principle of Least Action supplies a foundation certainly wide enough, if only it is not beyond our powers of development.

We require first to construct a dynamical scheme for the free aether when no material molecules are present. It is of course an elastic medium: let us assume that it is practically at rest, and let the vector (ξ, η, ζ) represent the displacement, elastic and other, of its substance at the point (x, y, z) which arises from the strain existing in it. We assume (to be hereafter verified by the results of the analysis) for its kinetic energy T and its potential energy W the expressions

$$T = \tfrac{1}{2} A \int (\dot{\xi}^2 + \dot{\eta}^2 + \dot{\zeta}^2)\, d\tau$$

$$W = \tfrac{1}{2} B \int (f^2 + g^2 + h^2)\, d\tau$$

in which $\delta\tau$ denotes an element of volume, A and B are constants, the former a constant of inertia, the latter a modulus of elasticity, and in which (f, g, h) is a vector defined as regards its mode of change[†] by the relation

$$(\dot{f}, \dot{g}, \dot{h}) = \frac{1}{4\pi}\left(\frac{d\dot\zeta}{dy} - \frac{d\dot\eta}{dz},\ \frac{d\dot\xi}{dz} - \frac{d\dot\zeta}{dx},\ \frac{d\dot\eta}{dx} - \frac{d\dot\xi}{dy}\right), \quad \text{(I)}$$

[†] This allows for the permanent existence, independently of $(\dot\xi, \dot\eta, \dot\zeta)$, of the intrinsic aethereal displacement surrounding each electron. Cf. Appendix E.

where the 4π is inserted in order to conform to the ordinary electrical usage.

This definition makes

$$\frac{df}{dx}+\frac{dg}{dy}+\frac{dh}{dz} = 0,$$

so that (f, g, h) is a stream vector.

To obtain the dynamical equations of this medium, we have to develop the variational equation

$$\delta \int (T-W)\, dt = 0,$$

subject to the time of motion being unvaried.

Now

$$\delta \int T dt = A \int dt \int (\dot{\xi}\delta\dot{\xi}+\dot{\eta}\delta\dot{\eta}+\dot{\zeta}\delta\dot{\zeta})\, d\tau$$
$$= A \left| \int (\dot{\xi}\delta\xi+\dot{\eta}\delta\eta+\dot{\zeta}\delta\zeta)\, d\tau \right|_{t_1}^{t_2}$$
$$- A \int dt \int (\ddot{\xi}\delta\xi+\ddot{\eta}\delta\eta+\ddot{\zeta}\delta\zeta)\, d\tau.$$

Also

$$\delta W = \frac{B}{4\pi} \int \left\{ f\left(\frac{d\delta\zeta}{dy}-\frac{d\delta\eta}{dz}\right)+g\left(\frac{d\delta\xi}{dz}-\frac{d\delta\zeta}{dx}\right)+h\left(\frac{d\delta\eta}{dx}-\frac{d\delta\xi}{dy}\right) \right\} d\tau$$
$$= \frac{B}{4\pi} \int \{(ng-mh)\,\delta\xi+(lh-nf)\,\delta\eta+(mf-lg)\,\delta\zeta\}\, dS$$
$$+ \frac{B}{4\pi} \int \left\{ \left(\frac{dh}{dy}-\frac{dg}{dz}\right)\delta\xi + \left(\frac{df}{dz}-\frac{dh}{dx}\right)\delta\eta + \left(\frac{dg}{dx}-\frac{df}{dy}\right)\delta\zeta \right\} d\tau$$

where (l, m, n) is the direction vector of the element of boundary surface δS.

In these reductions by integration by parts the aim has been as usual to express dependent variations such as $\delta\xi$, $d\delta\zeta/dy$, in terms of the independent ones $\delta\xi$, $\delta\eta$, $\delta\zeta$. This requires the introduction of surface integrals: if the region under consideration is infinite space, and the exciting causes of the disturbance are all at finite distance from the origin, these surface integrals over an infinitely remote boundary cannot in the nature of things be of influence on the state of the system at a finite distance, and in fact it may be verified that they give a null result: in other cases they must of course be retained.

On substitution in the equation of Action of these expressions for the variations, the coefficients of $\delta\xi$, $\delta\eta$, $\delta\zeta$ must separately vanish both in the volume integral and in the surface integral, since $\delta\xi$, $\delta\eta$, $\delta\zeta$ are perfectly independent and arbitrary both at each element of volume $\delta\tau$ and at each element of surface δS. This gives, from the volume integral, the equations of vibration or wave-propagation

$$\frac{B}{4\pi}\left(\frac{dh}{dy}-\frac{dg}{dz},\ \frac{df}{dz}-\frac{dh}{dx},\ \frac{dg}{dx}-\frac{df}{dy}\right) = -A(\ddot{\xi},\ddot{\eta},\ddot{\zeta}). \qquad \text{(II)}$$

The systems of equations (I) and (II), thus arrived at, become identical in form with Maxwell's circuital equations which express the electrostatic and electrodynamic working of free aether, if (ξ, η, ζ) represents the magnetic induction and (f, g, h) the aethereal displacement; the velocity of propagation is $(4\pi)^{-1}(B/A)^{\frac{1}{2}}$, so that $B/A = 16\pi^2 c^2$ where c is the velocity of radiation. They are also identical with MacCullagh's optical equations, the investigation here given being in fact due to him.

51. Now let us extend the problem to aether containing a system of electrons or discrete electric charges. Each of these point-charges determines a field of electric force around it: electric force must involve aether-strain of some kind, as has already been explained: thus an electric point-charge is a nucleus of intrinsic strain in the aether. It is not at present necessary to determine what kind of permanent configuration of strain in the aether this can be, if only we are willing to admit that it can move or slip freely about through

that medium much in the way that a knot slips along a rope: we thus in fact treat an electron or point-charge of strength e as a freely mobile singular point in the specification of the aethereal strain (f, g, h), such that very near to it (f, g, h) assumes the form

$$-\frac{e}{4\pi}\left(\frac{d}{dx}, \frac{d}{dy}, \frac{d}{dz}\right)\frac{1}{r}.$$

We can avoid the absolutely infinite values, at the origin of the distance r, by treating the nucleus of the permanent strain-form not as a point but as a very minute region:[†] this analytical artifice will keep all the elements of the integrals of our analysis finite, while it will not affect any physical application which considers the electron simply as a local charge of electricity of definite amount.

Now provided there is nothing involved in the electron except a strain-form, no inertia or energy foreign to the aether residing in its nucleus such as would prevent free unresisted mobility, as it is perhaps difficult to see how there could be, the equations (I) and (II) still determine the state of the field of aether, at any instant, from its state, supposed completely known, at the previous instant: and this determination includes a knowledge of the displacement of the nucleus of each strain-form during the intervening element of time. These equations therefore suffice to trace the natural sequence of change in the complex medium thus constituted by the aether and the nuclei pervading it. But if the nuclei had inertia and mutual actions of their own, independent of the aether, there would in addition to the continuous equations of motion of the aether itself be dynamical equations of motion for each strain-form as well, which would interact and so have to be combined into continuity with the aethereal equations, and the problem would assume a much more complex form: in other words, the complete energy function employed in formulating the Principle of Least

[†] This substitution affects only the *intrinsic* molecular energy; cf. *Phil. Trans.* 1894 A, pp. 812–3.

Action would also involve these other types of physical action, if they existed.

52. But for purposes of the electrodynamic phenomena of material bodies, which we can only test by observation and experiment on matter in bulk, a complete atomic analysis of the kind thus indicated would (even if possible) be useless; for we are unable to take direct cognizance of a single molecule of matter, much less of the separate electrons in the molecule to which this analysis has regard. The development of the theory which is to be in line with experience must instead concern itself with an effective differential element of volume, containing a crowd of molecules numerous enough to be expressible continuously, as regards their average relations, as a volume-density of matter. As regards the actual distribution in the element of volume of the really discrete electrons, all that we can usually take cognizance of is an excess of one kind, positive or negative, which constitutes a volume density of electrification, or else an average polarization in the arrangement of the groups of electrons in the molecules which must be specified as a vector by its intensity per unit volume: while the movements of the electrons, free and paired, in such element of volume must be combined into statistical aggregates of translational fluxes and molecular whirls of electrification. With anything else than mean aggregates of the various types that can be thus separated out, each extended over the effective element of volume, mechanical science, which has for its object matter in bulk as it presents itself to our observation and experiment, is not directly concerned: there is however another more abstract study, that of molecular dynamics, whose province it is to form and test hypotheses of molecular structure and arrangement, intended to account for the distinctive features of the mechanical phenomena aforesaid.

As the integral $\int (lf + mg + nh)\, dS$, extended over the boundary of any region, no longer vanishes when there are electrons in that region, it follows that the vector (f, g, h) which represents the strain or "electric displacement" of the aether, is no longer circuital when these individual electrons are merged in volume-densities, as they are when we consider a material medium continuously distributed, instead of merely the aether existing between its molecules; thus the definition of the mode of change of aethereal elastic displacement namely

$$4\pi(\dot{f}, \dot{g}, \dot{h}) = \operatorname{curl}(\xi, \eta, \zeta),$$

which held for free aether, would now be a contradiction in terms. In order to ascertain what is to replace this definition, let us consider the translation of a single electron e from a point P_1 to a neighbouring point P_2. This will cause an addition to the elastic strain (f, g, h) of the aether, represented by a strain-vector distributed with reference to lines which begin at P_1 and end at P_2, the addition being in fact the electric displacement due to the doublet formed by $-e$ at P_1 and $+e$ at P_2. This additional flux of electric displacement from P_2 to P_1 along these lines is not by itself circuital; but the circuits of the flux will be completed if we add to it a linear flux of electricity of the same total amount e, back again from P_1 to P_2 along the line P_1P_2. If we complete in this way the fluxes of *aethereal electric displacement*, due to the changes of position of all the electrons of the system, by the fluxes of these *true electric charges* through the aether, a new vector is obtained which we may call the flux of the *total electric displacement* per unit volume; and this vector forms a fundamentally useful conception from the circumstance that it is everywhere and always a circuital or stream vector.

We may now express this result analytically: to the rate of change of aethereal displacement $(\dot{f}, \dot{g}, \dot{h})\,\delta\tau$ in the element of volume $\delta\tau$ there must be added $\Sigma(e\dot{x}, e\dot{y}, e\dot{z})$, where $(\dot{x}, \dot{y}, \dot{z})$ is the velocity of a contained electron e, in order to get a circuital result: the *current of aethereal electric displacement* by itself is not circuital when averaged with regard to this element of volume, but the so-called *total current*, made up of it, and of the *true electric current* formed by the moving electrons, possesses that property.

Thus we have to deal, in the mechanical theory, with a more complex problem: instead of only aethereal displacement we have now two *independent* variables, aethereal displacement, and true electric current or flux of electrons. In the molecular analysis, on the other hand, the minute knowledge of aethereal displacement between and around the electrons of the molecules involved that of the movements of these electrons or singularities themselves, and there was only one independent variable, at any rate when the singularities are purely aethereal. The transition, from the complete knowledge of aether and individual molecules to the averaged and smoothed out specification of the element of volume of the complex medium, requires the presence of two independent variables, one for the aether and one for the matter, instead of a single variable only.

53. We may consider this fundamental explanation from a different aspect. There are present in the medium electrons or electric charges each of amount e, so that for any region Faraday's hypothesis gives

$$\int (lf+mg+nh)\, dS = \Sigma e;$$

and therefore, any finite change of state being denoted by $\triangle$, $\triangle \int (lf+mg+nh)\, dS$ is equal to the flux of electrons into the region across the boundary. Thus for example

$$\frac{d}{dt} \int (lf+mg+nh)\, dS = -\int (lu_0+mv_0+nw_0)\, dS$$

in which (u_0, v_0, w_0) is the true electric current which is simply this flux of electrons reckoned per unit time: hence transposing all the terms to the same side, we have for any closed surface

$$\int (lu+mv+nw)\, dS = 0,$$

where $(u, v, w) = (df/dt+u_0,\ dg/dt+v_0,\ dh/dt+w_0)$.

This relation expresses that (u, v, w), the total current of Maxwell's theory, is circuital or a stream.

The true current (u_0, v_0, w_0) above defined includes all the possible types of co-ordinated or averaged motions of electrons, namely, currents arising from conduction, from material polarization and its convection, from convection of charged bodies.

54. We have now to fix the meaning to be attached to $(\xi, \dot{\eta}, \zeta)$ or (a, b, c) in a mechanical theory which treats only of sensible elements of volume. Obviously it must be the mean value of this vector, as previously employed, for the aether in each element of volume. With this meaning it is now to be shown that the curl of $(\xi, \dot{\eta}, \zeta)$ is equal to $4\pi\,(u, v, w)$. We shall in fact see that for any open geometrical surface or sheet S of sensible extent, fixed in space, bounded by a contour s, Sir George Stokes' fundamental analytical theorem of transformation of a surface integral into a line integral round its contour, must under the present circumstances assume the wider form

$$\frac{1}{4\pi}\,\triangle \int \left(\xi\frac{dx}{ds}+\dot{\eta}\frac{dy}{ds}+\zeta\frac{dz}{ds}\right) ds = \triangle \int (l\dot{f}+m\dot{g}+n\dot{h})\, dS + \mathfrak{F} \quad \text{(i)}$$

where the symbol $\triangle$ represents the change in the integral which follows it, produced by the motion of the system in any finite time, and $\mathfrak{F}$ represents the total flux of electrons through the fixed surface S during that time. To this end consider two sheets S and S' both abutting on the same contour s: then as the two together form a closed surface we have

$$\int (l'f+m'g+n'h)\, dS' - \int (lf+mg+nh)\, dS = \Sigma e \quad \text{(ii)}$$

where Σe denotes the sum of the strengths of the electrons included between the sheets: in this formula the direction vectors (l', m', n') and (l, m, n) are both measured towards the same sides of the surfaces, which for the former S' is the side away from the region enclosed between them. Now if one of these included electrons moves across the surface S', the form of the integral for that surface will be abruptly altered, an element of it becoming infinite at the transition when the electron is on the surface; and this will vitiate the proof of Stokes' theorem considered as applying to the change in the value of that surface integral. But the form of the integral for the other surface, across which the electron has not penetrated, will not pass through any critical stage, and Stokes' theorem will still hold for the change caused in it. That is, for the latter surface the equation (i) will hold good in the ordinary way without any term such as $\mathfrak{F}$; and therefore by (ii), for the former surface, across which electrons are taken to pass, the term $\mathfrak{F}$ as above is involved.

The relation of Sir George Stokes, thus generalized, in which $\mathfrak{F}$ represents the total flux of electrons across the surface S, leads directly to the equation

$$\operatorname{curl}\,(\xi,\,\dot{\eta},\,\zeta) = 4\pi(\dot{f}+u_0,\,\dot{g}+v_0,\,\dot{h}+w_0)$$

where the vectors *now* represent mean values throughout the element of volume.

This relation holds, whether the system of molecules contained in the medium is *magnetically* polarized or not, for the transference of magnetic polarity across the sheet S cannot add anything to the electric flux through it: it appears therefore that in a case involving magnetic polarization $(\dot{\xi},\,\dot{\eta},\,\dot{\zeta})$ represents what is called the magnetic induction and not the magnetic force, which is also in keeping with the stream character of the former vector. On the other hand the change in the *electric* polarization $(f',\,g',\,h')$ of the molecules constitutes an addition $\triangle(f',\,g',\,h')$ of finite amount per unit area to the flux through the sheet, so that $d/dt\,(f',\,g',\,h')$ constitutes a part of the true electric current $(u_0,\,v_0,\,w_0)$.

. . . . .

Chapter X

GENERAL PROBLEM OF MOVING MATTER TREATED IN RELATION TO THE INDIVIDUAL MOLECULES

Formulation of the Problem

102. WE shall now consider the material system as consisting of free aether pervaded by a system of electrons which are to be treated individually, some of them free or isolated, but the great majority of them grouped into material molecules: and we shall attempt to compare the relative motions of these electrons when they form, or belong to, a material system devoid of translatory motion through the aether, with what it would be when a translatory velocity is superposed, say for shortness a velocity v parallel to the axis of x. The medium in which the activity occurs is for our present purpose the free aether itself, whose dynamical equations have been definitely ascertained in quite independent ways from consideration of both the optical side and the electrodynamic side of its activity: so that there will be nothing hypothetical in our analysis on that score. An electron e will occur in this analysis as a singular point in the aether, on approaching which the elastic strain constituting the aethereal displacement (f, g, h) increases indefinitely, according to the type

$$-e/4\pi \cdot (d/dx, \quad d/dy, \quad d/dz)r^{-1}:$$

it is in fact analogous to what is called a simple pole in the two-dimensional representation that is employed in the theory of a function of a complex variable. It is assumed that this singularity represents a definite structure, forming a nucleus of strain in the aether, which is capable of transference across that medium independently of motion of the aether itself: the portion of the surrounding aethereal strain, of which the displacement-vector (f, g, h)

is the expression, which is associated with the electron and is carried along with the electron in its motion, being as above $-e/4\pi \times (d/dx, d/dy, d/dz)r^{-1}$. It is to be noticed that the energy of this part of the displacement is closely concentrated around the nucleus of the electron, and not widely diffused as might at first sight appear. The aethereal displacement satisfies the stream-condition

$$df/dx + dg/dy + dh/dz = 0,$$

except where there are electrons in the effective element of volume: these are analogous to the so-called sources and sinks in the abstract theory of liquid flow, so that when electrons are present the integral of the normal component of the aethereal displacement over the boundary of any region, instead of being null, is equal to the quantity Σe of electrons existing in the region. The other vector which is associated with the aether, namely the magnetic induction (a, b, c), also possesses the stream property; but singular points in its distribution, of the nature of simple poles, do not exist. The motion of an electron involves however a singularity in (a, b, c), of a rotational type, with its nucleus at the moving electron;† and the time-average of this singularity for a very rapid minute steady

---

† Namely as the distance r from it diminishes indefinitely, the magnetic induction tends to the form $evr^{-2} \sin \theta$, at right angles to the plane of the angle θ between r and the velocity v of the electron: this arises as the disturbance of the medium involved in annulling the electron in its original position and restoring it in the new position to which it has moved. The relations will appear more clearly when visualized by the kinematic representation of Appendix E; or when we pass to the limit in the formulae of Chapter ix relating to the field of a moving charged body of finite dimensions.

The specification in the text, as a simple pole, only applies for an electron moving with velocity v, when terms of the order $(v/c)^2$ are neglected: otherwise the aethereal field close around it is not isotropic and an amended specification derivable from the formulae of Chapter ix must be substituted. In the second-order discussion of Chapter xi this more exact form is implicitly involved, the strength of the electron being determined (§ 111) by the concentration of the aethereal displacement around it. The singularity in the magnetic field which is involved in the motion of the electron, not of course an intrinsic one, has no concentration.

orbital motion of an electron is analytically equivalent, at distances considerable compared with the dimensions of the orbit, to a magnetic doublet analogous to a source and associated equal sink. Finally, the various parts of the aether are supposed to be sensibly at rest, so that for example the time-rate of change of the strain of any element of the aether is represented by differentiation with respect to the time without any additional terms to represent the change due to the element of aether being carried on in the meantime to a new position; in this respect the equations of the aether are much simpler than those of the dynamics of fluid motion, being in fact linear. The aether is stagnant on this theory, while the molecules constituting the Earth and all other material bodies flit through it without producing any finite flow in it; hence the law of the astronomical aberration of light is rigorously maintained, and the Doppler change of wave-length of radiation from a moving source holds good; but it will appear that all purely terrestrial optical phenomena are unaffected by the Earth's motion.

103. Subject to this general explanation, the analytical equations which express the dynamics of the field of free aether, existing between and around the nuclei of the electrons, are

$$4\pi \frac{d}{dt}(f, g, h) = \text{curl}(a, b, c)$$

$$-\frac{d}{dt}(a, b, c) = 4\pi c^2 \text{curl}(f, g, h),$$

in which the symbol curl (a, b, c) represents, after Maxwell, the vector

$$\left(\frac{dc}{dy} - \frac{db}{dz}, \frac{da}{dz} - \frac{dc}{dx}, \frac{db}{dx} - \frac{da}{dy}\right),$$

and in which c is the single physical constant of the aether, being the velocity of propagation of elastic disturbances through it. These are the analytical equations derived by Maxwell in his mathematical development of Faraday's views as to an electric medium: and they are the same as the equations arrived at by MacCullagh a quarter of a century earlier in his formulation of the dynamics of optical media. It may fairly be claimed that the theoretical investigations of Maxwell, in combination with the experimental verifications of Hertz and his successors in that field, have imparted to this analytical formulation of the dynamical relations of free aether an exactness and precision which is not surpassed in any other department of physics, even in the theory of gravitation.

Where a more speculative element enters is in the construction of a kinematic scheme of representation of the aether-strain, such as will allow of the unification of the various assumptions here enumerated. It is desirable for the sake of further insight, and even necessary for various applications, to have concrete notions of the physical nature of the vectors (f, g, h) and (a, b, c) which specify aethereal disturbances, in the form of representations such as will implicitly and intuitively involve the analytical relations between them, and will also involve the conditions and restrictions to which each is subject, including therein the permanence and characteristic properties of an electron and its free mobility through the aether.[†]

104. But for the mere analytical development of the aether-scheme as above formulated, a concrete physical representation of the constitution of the aether is not required: the abstract relations and conditions above given form a sufficient basis. In point of fact these analytical relations are theoretically of an ideal simplicity for this purpose: for they give explicitly the time-rates of change of the vectors of the problem at each instant, so that from a knowledge of the state of the system at any time t the state at the time $t + \delta t$ can be immediately expressed, and so by successive steps, or by the use of Taylor's differential expansion-theorem, its state at any further time can theoretically be derived. The point that requires careful attention is as to whether the solution of these

---

[†] See Appendix E.

equations in terms of a given initial state of the system determines the motions of the electrons or strain-nuclei through the medium, as well as the changes of strain in the medium itself: and it will appear on consideration that under suitable hypotheses this is so. For the given initial state will involve given motions of the electrons, that is the initial value of (a, b, c) will involve rotational singularities at the electrons around their directions of motion, just such as in the element of time δt will shift the electrons themselves into their new positions:† and so on step by step continually. This however presupposes that the nucleus of the electron is quite labile as regards displacement through the aether, in other words that its movement is not influenced by any inertia or forces except such as are the expression of its relation to the aether: we in fact assume the *completeness* of the aethereal scheme of relations as above given. Any difficulty that may be felt on account of the infinite values of the vectors at the nucleus itself may be removed, in the manner customary in analytical discussions on attractions, by considering the nucleus to consist of a volume distribution of electricity of finite but very great density, distributed through a very small space instead of being absolutely concentrated in a point: then the quantities will not become infinite. Of the detailed structure of electrons nothing is assumed: so long as the actual dimensions of their nuclei are extremely small in comparison with the distances between them, it will suffice for the theory to consider them as points, just as for example in the general gravitational theory of the Solar System it suffices to consider the planets as attracting points. This method is incomplete only as regards those portions of the energy and other quantities that are associated with the mutual actions of the parts of the electron itself, and are thus molecularly constitutive.

105. It is to be observed that on the view here being developed, in which atoms of matter are constituted of aggregations of elec-

† Cf. footnote, p. 162. [p. 225 in this version — K.F.S.]

trons, the only actions between atoms are what may be described as electric forces. The electric character of the forces of chemical affinity was an accepted part of the chemical views of Davy, Berzelius, and Faraday; and more recent discussions, while clearing away crude conceptions, have invariably tended to the strengthening of that hypothesis. The mode in which the ordinary forces of cohesion could be included in such a view is still quite undeveloped. Difficulties of this kind have however not been felt to be fundamental in the vortex-atom illustration of the constitution of matter, which has exercised much fascination over high authorities on molecular physics: yet in the concrete realization of Maxwell's theory of the aether above referred to, the atom of matter possesses all the dynamical properties of a vortex ring in a frictionless fluid, so that everything that can be done in the domain of vortex-ring illustration is implicitly attached to the present scheme. The fact that virtually nothing has been achieved in the department of forces of cohesion is not a valid objection to the development of a theory of the present kind. For the aim of theoretical physics is not a complete and summary conquest of the *modus operandi* of natural phenomena: that would be hopelessly unattainable if only for the reason that the mental apparatus with which we conduct the search is itself in one of its aspects a part of the scheme of Nature which it attempts to unravel. But the very fact that this is so is evidence of a correlation between the process of thought and the processes of external phenomena, and is an incitement to push on further and bring out into still clearer and more direct view their interconnexions. When we have mentally reduced to their simple elements the correlations of a large domain of physical phenomena, an objection does not lie because we do not know the way to push the same principles to the explanation of other phenomena to which they should presumably apply, but which are mainly beyond the reach of our direct examination.

The natural conclusion would rather be that a scheme, which has been successful in the simple and large-scale physical phenomena that we can explore in detail, must also have its place, with proper modifications or additions on account of the difference of scale, in the more minute features of the material world as to which direct knowledge in detail is not available. And in any case, whatever view may be held as to the necessity of the whole complex of chemical reaction being explicable in detail by an efficient physical scheme, a limit is imposed when vital activity is approached: any complete analysis of the conditions of the latter, when merely superficial sequences of phenomena are excluded, must remain outside the limits of our reasoning faculties. The object of scientific explanation is in fact to coordinate mentally, but not to exhaust, the interlaced maze of natural phenomena: a theory which gives an adequate correlation of a portion of this field maintains its place until it is proved to be in definite contradiction, not removable by suitable modification, with another portion of it.

Application to moving Material Media: approximation up to first order

106. We now recall the equations of the free aether, with a view to changing from axes (x, y, z) at rest in the aether to axes (x', y', z') moving with translatory velocity v parallel to the axis of x; so as thereby to be in a position to examine how phenomena are altered when the observer and his apparatus are in uniform motion through the stationary aether. These equations are

$$4\pi \frac{df}{dt} = \frac{dc}{dy} - \frac{db}{dz} \qquad -(4\pi c^2)^{-1} \frac{da}{dt} = \frac{dh}{dy} - \frac{dg}{dz}$$

$$4\pi \frac{dg}{dt} = \frac{da}{dz} - \frac{dc}{dx} \qquad -(4\pi c^2)^{-1} \frac{db}{dt} = \frac{df}{dz} - \frac{dh}{dx}$$

$$4\pi \frac{dh}{dt} = \frac{db}{dx} - \frac{da}{dy} \qquad -(4\pi c^2)^{-1} \frac{dh}{dt} = \frac{dg}{dx} - \frac{df}{dy}.$$

When they are referred to the axes (x', y', z') in uniform motion, so that $(x', y', z') = (x - vt, y, z)$, $t' = t$, then d/dx, d/dy, d/dz become

d/dx', d/dy', d/dz', but d/dt becomes $d/dt' - v d/dx'$: thus

$$4\pi \frac{df}{dt'} = \frac{dc'}{dy'} - \frac{db'}{dz'} \qquad -(4\pi c^2)^{-1} \frac{da}{dt'} = \frac{dh'}{dy'} - \frac{dg'}{dz'}$$

$$4\pi \frac{dg}{dt'} = \frac{da'}{dz'} - \frac{dc'}{dx'} \qquad -(4\pi c^2)^{-1} \frac{db}{dt'} = \frac{df'}{dz'} - \frac{dh'}{dx'}$$

$$4\pi \frac{dh}{dt'} = \frac{db'}{dx'} - \frac{da'}{dy'} \qquad -(4\pi c^2)^{-1} \frac{dc}{dt'} = \frac{dg'}{dx'} - \frac{df'}{dy'},$$

where

$$(a', b', c') = (a, b + 4\pi vh, c - 4\pi vg)$$
$$(f', g', h') = \left(f, g - \frac{v}{4\pi c^2} c, h + \frac{v}{4\pi c^2} b\right).$$

We can complete the elimination of (f, g, h) and (a, b, c) so that only the vectors denoted by accented symbols shall remain, by substituting from these latter formulae: thus

$$g = g' + \frac{v}{4\pi c^2} (c' + 4\pi vg),$$

so that $\qquad \varepsilon^{-1} g = g' + \dfrac{v}{4\pi c^2} c',$

where ε is equal to $(1 - v^2/c^2)^{-1}$, and exceeds unity;

and $\qquad b = b' - 4\pi v\left(h' - \dfrac{v}{4\pi c^2} b\right)$

so that $\qquad \varepsilon^{-1} b = b' - 4\pi vh';$

giving the general relations

$$\varepsilon^{-1}(a, b, c) = (\varepsilon^{-1} a', b' - 4\pi vh', c' + 4\pi vg')$$
$$\varepsilon^{-1}(f, g, h) = \left(\varepsilon^{-1} f', g' + \frac{v}{4\pi c^2} c', h' - \frac{v}{4\pi c^2} b'\right).$$

Hence

$$4\pi \frac{df'}{dt'} = \frac{dc'}{dy'} - \frac{db'}{dz'}$$

$$4\pi\varepsilon \frac{dg'}{dt'} = \frac{da'}{dz'} - \left(\frac{d}{dx'} + \frac{v}{c^2}\varepsilon\frac{d}{dt'}\right)c'$$

$$4\pi\varepsilon \frac{dh'}{dt'} = \left(\frac{d}{dx'} + \frac{v}{c^2}\varepsilon\frac{d}{dt'}\right)b' - \frac{da'}{dy'}$$

$$-(4\pi c^2)^{-1} \frac{da'}{dt'} = \frac{dh'}{dy'} - \frac{dg'}{dz'}$$

$$-(4\pi c^2)^{-1}\varepsilon \frac{db'}{dt'} = \frac{df'}{dz'} - \left(\frac{d}{dx'} + \frac{v}{c^2}\varepsilon\frac{d}{dt'}\right)h'$$

$$-(4\pi c^2)^{-1}\varepsilon \frac{dc'}{dt'} = \left(\frac{d}{dx'} + \frac{v}{c^2}\varepsilon\frac{d}{dt'}\right)g' - \frac{df'}{dy'}.$$

Now change the time-variable from t' to t'', equal to $t'-(v/c^2)\varepsilon x'$; this will involve that

$$\frac{d}{dx'} + \frac{v}{c^2}\varepsilon\frac{d}{dt'}$$

is replaced by d/dx', while the other differential operators remain unmodified; thus the scheme of equations reverts to the same type as when it was referred to axes at rest, except as regards the factors ε on the left-hand sides.

107. It is to be observed that this factor ε only differs from unity by $(v/c)^2$, which is of the second order of small quantities; hence we have the following correspondence when that order is neglected. Consider any aethereal system, and let the sequence of its spontaneous changes referred to axes (x', y', z') moving uniformly through the aether with velocity $(v, 0, 0)$ be represented by values of the vectors (f, g, h) and (a, b, c) expressed as functions of x', y', z' and t', the latter being the time measured in the ordinary manner: then there exists a correlated aethereal system whose sequence of spontaneous changes referred to axes (x', y', z') at rest are such

that its electric and magnetic vectors (f', g', h') and (a', b', c') are functions of the variables x', y', z' and a time-variable t'', equal to $t' - (v/c^2)x'$, which are the same as represent the quantities

$$\left(f, g - \frac{v}{4\pi c^2} c, h + \frac{v}{4\pi c^2} b\right).$$

and $(a, b + 4\pi v h, c - 4\pi v g)$

belonging to the related moving system when expressed as functions of the variables x', y', z' and t'.

Conversely, taking any aethereal system at rest in the aether, let the sequence of its changes be represented by (f', g', h') and (a', b', c') expressed as functions of the co-ordinates (x, y, z) and of the time t'. In these functions change t' into $t - (v/c^2)x$: then the resulting expressions are the values of

$$\left(f, g - \frac{v}{4\pi c^2} c, h + \frac{v}{4\pi c^2} b\right),$$

and $(a, b + 4\pi v h, c - 4\pi v g)$,

for a system in uniform motion through the aether, referred to axes (x, y, z) moving along with it, and to the time t. In comparing the states of the two systems, we have to the first order

$$\frac{df}{dx} \quad \text{equal to} \quad \frac{df'}{dx} - \frac{v}{c^2} \frac{df'}{dt}$$

$$\frac{d}{dy}\left(g - \frac{v}{4\pi c^2} c\right) \quad \text{equal to} \quad \frac{dg'}{dy}$$

$$\frac{d}{dz}\left(h + \frac{v}{4\pi c^2} b\right) \quad \text{equal to} \quad \frac{dh'}{dz};$$

hence bearing in mind that for the system at rest

$$\frac{dc'}{dy} - \frac{db'}{dz} = 4\pi \frac{df'}{dt'},$$

or, what is the same,

$$\frac{dc'}{dy} - \frac{db'}{dz} = 4\pi\left(\frac{df'}{dt} - v\frac{df'}{dx}\right),$$

we have, to the first order,

$$\frac{df}{dx} + \frac{dg}{dy} + \frac{dh}{dz} = \frac{df'}{dx} + \frac{dg'}{dy} + \frac{dh'}{dz}.$$

Thus the electrons in the two systems here compared, being situated at the singular points at which the concentration of the electric displacement ceases to vanish, occupy corresponding positions. Again, these electrons are of equal strengths: for, very near an electron, fixed or moving, the values of (f, g, h) and (a, b, c) are practically those due to it, the part due to the remainder of the system being negligible in comparison: also in this correspondence the relation between (f, g, h) and the accented variables is, by § 106

$$\varepsilon^{-1}(f, g, h) = \left(\varepsilon^{-1}f', g' + \frac{v}{4\pi c^2}c', h' - \frac{v}{4\pi c^2}b'\right);$$

hence, since for the single electron at rest (a', b', c') is null, we have, very close to the correlative electron in the moving system, (f, g, h) equal to $(f', \varepsilon g', \varepsilon h')$, where ε, being $(1 - v^2/c^2)^{-1}$, differs from unity by the second order of small quantities. Thus neglecting the second order, (f, g, h) is equal to (f', g', h') for corresponding points very close to electrons; and, as the amount of electricity inside any boundary is equal to the integral of the normal component of the aethereal displacement taken over the boundary, it follows by taking a very contracted boundary that the strengths of the corresponding electrons in the two systems are the same, to this order of approximation.

108. It is to be observed that the above analytical transformation of the equations applies to any isotropic dielectric medium as well as to free aether: we have only to alter c into the velocity of radia-

tion in that medium, and all will be as above. The transformation will thus be different for different media. But we are arrested if we attempt to proceed to compare a moving material system, treated as continuous, with the same system at rest; for the motion of the polarized dielectric matter has altered the mathematical type of the electric current. It is thus of no avail to try to effect in this way a direct general transformation of equations of a material medium in which dielectric and conductive coefficients occur.

109. The correspondence here established between a system referred to fixed axes and a system referred to moving axes will assume a very simple aspect when the former system is a steady one, so that the variables are independent of the time. Then the distribution of electrons in the second system will be at each instant precisely the same as that in the first, while the second system accompanies its axes of reference in their uniform motion through the aether. In other words, given any system of electrified bodies at rest, in equilibrium under their mutual electric influences and imposed constraints, there will be a precisely identical system in equilibrium under the same constraints, and in uniform translatory motion through the aether. That is, uniform translatory motion through the aether does not produce any alteration in electric distributions as far as the first order of the ratio of the velocity of the system to the velocity of radiation is concerned. Various cases of this general proposition will be verified subsequently in connexion with special investigations.

Moreover this result is independent of any theory as to the nature of the forces between material molecules: the structure of the matter being assumed unaltered to the first order by motion through the aether, so too must be all electric distributions. What has been proved comes to this, that if any configuration of ionic charges is the natural one in a material system at rest, the maintenance of the same configuration as regards the system in uniform motion will not require the aid of any new forces. The electron *taken by itself* must be on any conceivable theory a simple singularity of the aether whose movements when it is free, and interactions with other electrons if it can be constrained by matter, are traceable through the differential equations of the surrounding free aether alone: and a correlation has been established between these equations for the two cases above compared. It is however to be observed (cf. § 99) that though the fixed and the moving system of electrons of this correlation are at corresponding instants identical, yet the electric and magnetic displacements belonging to them differ by terms of the first order.

Chapter XI

MOVING MATERIAL SYSTEM: APPROXIMATION CARRIED TO THE SECOND ORDER

110. The results above obtained have been derived from the correlation developed in § 106, up to the first order of the small quantity v/c, between the equations for aethereal vectors here represented by (f', g', h') and (a', b', c') referred to the axes (x', y', z') at rest in the aether and a time t'', and those for related aethereal vectors represented by (f, g, h) and (a, b, c) referred to axes (x', y', z') in uniform translatory motion and a time t'. But we can proceed further, and by aid of a more complete transformation institute a correspondence which will be correct to the second order. Writing as before t'' for $t' - (v/c^2)\varepsilon x'$, the exact equations for (f, g, h) and (a, b, c) referred to the moving axes (x', y', z') and time t' are, as above shown, equivalent to

$$4\pi \frac{df'}{dt''} = \frac{dc'}{dy'} - \frac{db'}{dz'} \qquad -(4\pi c^2)^{-1} \frac{da'}{dt''} = \frac{dh'}{dy'} - \frac{dg'}{dz'}$$

$$4\pi\varepsilon \frac{dg'}{dt''} = \frac{da'}{dz'} - \frac{dc'}{dx'} \qquad -(4\pi c^2)^{-1}\varepsilon \frac{db'}{dt''} = \frac{df'}{dz'} - \frac{dh'}{dx'}$$

$$4\pi\varepsilon \frac{dh'}{dt''} = \frac{db'}{dx'} - \frac{da'}{dy'} \qquad -(4\pi c^2)^{-1}\varepsilon \frac{dc'}{dt''} = \frac{dg'}{dx'} - \frac{df'}{dy'}.$$

Now write

(x_1, y_1, z_1) for $\left(\varepsilon^{\frac{1}{2}} x', y', z'\right)$

(a_1, b_1, c_1) for $\left(\varepsilon^{-\frac{1}{2}} a', b', c'\right)$ or $\left(\varepsilon^{-\frac{1}{2}} a, +4\pi vh, c - 4\pi vg\right)$

(f_1, g_1, h_1) for $\left(\varepsilon^{-\frac{1}{2}} f', g', h'\right)$

or $\left(\varepsilon^{-\frac{1}{2}} f, g - \dfrac{v}{4\pi c^2} c, h + \dfrac{v}{4\pi c^2} b\right)$

dt_1 for $\varepsilon^{-\frac{1}{2}} dt''$ or $\varepsilon^{-\frac{1}{2}}\left(dt' - \dfrac{v}{c^2} \varepsilon dx'\right),$

where $\varepsilon = (1 - v^2/c^2)^{-1}$; and it will be seen that the factor ε is absorbed, so that the scheme of equations, referred to moving axes, which connects together the new variables with subscripts, is identical in form with the Maxwellian scheme of relations for the aethereal vectors referred to fixed axes. This transformation, from (x', y', z') to (x_1, y_1, z_1) as dependent variables, signifies an elongation of the space of the problem in the ratio $\varepsilon^{\frac{1}{2}}$ along the direction of the motion of the axes of coordinates. Thus if the values of (f_1, g_1, h_1) and (a_1, b_1, c_1) given as functions of x_1, y_1, z_1, t_1 express the course of spontaneous change of the aethereal vectors of a system of moving electrons referred to axes (x_1, y_1, z_1) at rest in the aether, then

$$\left(\varepsilon^{-\frac{1}{2}} f, g - \frac{v}{4\pi c^2} c, h + \frac{v}{4\pi c^2} b\right)$$

and $\left(\varepsilon^{-\frac{1}{2}} a, b + 4\pi vh, c - 4\pi vg\right),$

expressed by the same functions of the variables

$$\varepsilon^{\frac{1}{2}} x', y', z', \varepsilon^{-\frac{1}{2}} t' - \frac{v}{c^2} \varepsilon^{\frac{1}{2}} x',$$

will represent the course of change of the aethereal vectors (f, g, h) and (a, b, c) of a correlated system of moving electrons referred to axes of (x', y', z') moving through the aether with uniform translatory velocity $(v, 0, 0)$. In this correlation between the courses of change of the two systems, we have

$$\frac{d(\varepsilon^{-\frac{1}{2}}f)}{d(\varepsilon^{\frac{1}{2}}x')} \quad \text{equal to} \quad \frac{df_1}{dx_1} - \frac{v}{c^2}\frac{df_1}{dt_1},$$

$$\frac{d}{dy'}\left(g - \frac{v}{4\pi c^2}c\right) \quad \text{equal to} \quad \frac{dg_1}{dy_1}$$

$$\frac{d}{dz'}\left(h + \frac{v}{4\pi c^2}b\right) \quad \text{equal to} \quad \frac{dh_1}{dz_1},$$

where
$$\frac{dc}{dy'} - \frac{db}{dz'} = 4\pi\left(\frac{df}{dt'} - v\frac{df}{dx'}\right)$$

and also
$$\frac{df_1}{dt_1} = \frac{df}{dt};$$

hence

$$\frac{df}{dx'} + \frac{dg}{dy'} + \frac{dh}{dz'} - \frac{v}{c^2}\left(\frac{df}{dt'} - v\frac{df}{dx'}\right)$$

is equal to
$$\varepsilon\frac{df_1}{dx_1} + \frac{dg_1}{dy_1} + \frac{dh_1}{dz_1} - \frac{v}{c^2}\varepsilon\frac{df}{dt},$$

so that, up to the order of $(v/c)^2$ inclusive,

$$\frac{df}{dx'} + \frac{dg}{dy'} + \frac{dh}{dz'} = \frac{df_1}{dx_1} + \frac{dg_1}{dy_1} + \frac{dh_1}{dz_1}.$$

Thus the conclusions as to the corresponding positions of the electrons of the two systems, which had been previously established up to the first order of v/c, are true up to the second order when the dimensions of the moving system are contracted in comparison with the fixed system in the ratio $\varepsilon^{-\frac{1}{2}}$, or $1 - \frac{1}{2}v^2/c^2$, along the direction of its motion.

111. The ratio of the strengths of corresponding electrons in the two systems may now be deduced just as it was previously when the discussion was confined to the first order of v/c. For the case of a single electron in uniform motion the comparison is with a single electron at rest, near which (a_1, b_1, c_1) vanishes so far as it depends on that electron: now we have in the general correlation

$$g = g_1 + \frac{v}{4\pi c^2}(c_1 + 4\pi v g),$$

hence in this particular case

$$(g, h) = \varepsilon(g_1, h_1), \quad \text{while} \quad f = \varepsilon^{\frac{1}{2}} f_1.$$

But the strength of the electron in the moving system is the value of the integral

$$\iint (f \, dy' \, dz' + g \, dz' \, dx' + h \, dx' \, dy')$$

extended over any surface closely surrounding its nucleus; that is here

$$\varepsilon^{\frac{1}{2}} \iint (f_1 \, dy_1 \, dz_1 + g_1 \, dz_1 \, dx_1 + h_1 \, dx_1 \, dy_1),$$

so that the strength of each moving electron is $\varepsilon^{\frac{1}{2}}$ times that of the correlative fixed electron. As before, no matter what other electrons are present, this argument still applies if the surface be taken to surround the electron under consideration very closely, because then the wholly preponderating part of each vector is that which belongs to the adjacent electron.[†]

112. We require however to construct a correlative system devoid of the translatory motion in which the strengths of the elect-

[†] This result follows more immediately from § 110, which shows that corresponding densities of electrification are equal, while corresponding volumes are as $\varepsilon^{\frac{1}{2}}$ to unity.

rons shall be equal instead of proportional, since motion of a material system containing electrons cannot alter their strengths. The principle of dynamical similarity will effect this.

We have in fact to reduce the scale of the electric charges, and therefore of

$$\frac{df}{dx} + \frac{dg}{dy} + \frac{dh}{dz},$$

in a system at rest in the ratio $\varepsilon^{-\frac{1}{2}}$. Apply therefore a transformation

$$(x, y, z) = k(x_1, y_1, z_1), \quad t = lt_1,$$
$$(a, b, c) = \vartheta(a_1, b_1, c_1), \quad (f, g, h) = \varepsilon^{-\frac{1}{2}}k(f_1, g_1, h_1);$$

and the form of the fundamental circuital aethereal relations will not be changed provided $k = l$ and $\vartheta = \varepsilon^{-\frac{1}{2}}k$. Thus we may have k and l both unity and $\vartheta = \varepsilon^{-\frac{1}{2}}$; so that no further change of scale in space and time is required, but only a diminution of (a, b, c) in the ratio $\varepsilon^{-\frac{1}{2}}$.

We derive the result, correct to the second order, that if the internal forces of a material system arise wholly from electrodynamic actions between the systems of electrons which constitute the atoms, then an effect of imparting to a steady material system a uniform velocity of translation is to produce a uniform contraction of the system in the direction of the motion, of amount $\varepsilon^{-\frac{1}{2}}$ or $1 - \frac{1}{2}v^2/c^2$. The electrons will occupy corresponding positions in this contracted system, but the aethereal displacements in the space around them will not correspond: if (f, g, h) and (a, b, c) are those of the moving system, then the electric and magnetic displacements at corresponding points of the fixed systems will be the values that the vectors

$$\varepsilon^{\frac{1}{2}}\left(\varepsilon^{-\frac{1}{2}}f, \ g - \frac{v}{4\pi c^2}c, \ h + \frac{v}{4\pi c^2}b\right)$$

and
$$\varepsilon^{\frac{1}{2}}\left(\varepsilon^{-\frac{1}{2}}a,\ b+4\pi vh,\ c-4\pi vg\right)$$

had at a time const.$+vx/c^2$ before the instant considered when the scale of time is enlarged in the ratio $\varepsilon^{\frac{1}{2}}$.

As both the electric and magnetic vectors of radiation lie in the wave-front, it follows that in the two correlated systems, fixed and moving, the relative wave-fronts of radiation correspond, as also do the rays which are the paths of the radiant energy relative to the systems. The change of the time variable, in the comparison of radiations in the fixed and moving systems, involves the Doppler effect on the wave-length.

The Correlation between a stationary and a moving Medium, as regards trains of Radiation

113. Consider the aethereal displacement given by
$$(f_1, g_1, h_1) = (L, M, N)\, F(lx_1+my_1+nz_1-pt),$$
which belongs to a plane wave-train advancing, along the direction (l, m, n) with velocity V, or c/μ where μ is refractive index, equal to
$$p(l^2+m^2+n^2)^{-\frac{1}{2}},$$
in the material medium at rest referred to coordinates (x_1, y_1, z_1). In the corresponding wave-train relative to the same medium in motion specified by coordinates (x, y, z), and considered as shrunk in the above manner as a result of the motion, the vectors (f, g, h) and (a, b, c) satisfy the relation

$$\varepsilon^{\frac{1}{2}}\left(\varepsilon^{-\frac{1}{2}}f,\ g-\frac{v}{4\pi c^2}c,\ h+\frac{v}{4\pi c^2}b\right)$$
$$= (L, M, N)\, F\left\{l\varepsilon^{\frac{1}{2}}x+my+nz-p\varepsilon^{-\frac{1}{2}}\left(t-\frac{v}{c^2}\varepsilon x\right)\right\}$$
$$= (L, M, N)\, F\left\{\left(l\varepsilon^{\frac{1}{2}}+\frac{pv}{c^2}\varepsilon^{\frac{1}{2}}\right)x+my+nz-p\varepsilon^{-\frac{1}{2}}t\right\}.$$

As the wave-train in the medium at rest is one of transverse displacement, so that the vectors (f_1, g_1, h_1) and (a_1, b_1, c_1) are both in the wave-front, the same is therefore true for the vectors (f, g, h) and (a, b, c) in the correlative wave-train in the moving system, as was in fact to be anticipated from the circuital quality of these vectors: the direction vector of the front of the latter train is proportional to

$$\left(l\varepsilon^{\frac{1}{2}} + \frac{pv}{c^2}\varepsilon^{\frac{1}{2}}, m, n\right),$$

and its velocity of propagation is

$$p\varepsilon^{-\frac{1}{2}} \Big/ \left\{\left(l\varepsilon^{\frac{1}{2}} + \frac{pv}{c^2}\varepsilon^{\frac{1}{2}}\right)^2 + m^2 + n^2\right\}^{\frac{1}{2}}.$$

Thus, when the wave-train is travelling with velocity V along the direction of translation of the material medium, that is along the axis of x so that m and n are null, the velocity of the train relative to the moving medium is

$$V\varepsilon^{-1} \Big/ \left(1 + \frac{Vv}{c^2}\right),$$

which is, to the second order,

$$V\left(1 - \frac{v^2}{c^2}\right) \Big/ \left(1 + \frac{Vv}{c^2}\right) \quad \text{or} \quad V - \frac{v}{\mu^2} - \left(\frac{1}{\mu} - \frac{1}{\mu^3}\right)\frac{v^2}{c}.$$

The second term in this expression is the Fresnel effect, and the remaining term is its second order correction on our hypothesis which includes Michelson's negative result.

In the general correlation, the wave-length in the train of radiation relative to the moving material system differs from that in the corresponding train in the same system at rest by the factor

$$\left(1 + 2l\frac{pv}{c^2}\right)^{-\frac{1}{2}}, \quad \text{or} \quad 1 - lv/\mu c,$$

where l is the cosine of the inclination of the ray to the direction of v; it is thus shorter by a quantity of the first order, which represents the Doppler effect on wave-length because the period is the same up to that order.

When the wave-fronts relative to the moving medium are travelling in a direction making an angle θ', in the plane xy so that n is null, with the direction of motion of the medium, the velocity V' of the wave-train (of wave-length thus altered) relative to the medium is given by

$$\frac{\cos \theta'}{V'} = \frac{l\varepsilon}{p} + \frac{v\varepsilon}{c^2}, \quad \frac{\sin \theta'}{V'} = \frac{m\varepsilon^{\frac{1}{2}}}{p},$$

where $(l^2+m^2)/p^2 = V^{-2}$. Thus

$$\left(\frac{\varepsilon^{-1} \cos \theta'}{V'} - \frac{v}{c^2}\right)^2 + \frac{\varepsilon^{-1} \sin^2 \theta'}{V'^2} = \frac{1}{V^2},$$

so that neglecting $(v/c)^3$,

$$V' = V - \frac{v}{\mu^2} \cos \theta' - \frac{1}{2}(1-\mu^{-2})\frac{v^2}{\mu c}(1+3\cos^2 \theta'),$$

where $\mu = c/V$, of which the last term is the general form of the second order correction to Fresnel's expression. In free aether, for which μ is unity, this formula represents the velocity relative to the moving axes of an unaltered wave-train, as it ought to do.

As (f, g, h) and (a, b, c) are in the same phase in the free transparent aether, when one of them is null so is the other: hence in any experimental arrangement, regions where there is no disturbance in the one system correspond to regions where there is no disturbance in the other. As optical measurements are usually made by the null method of adjusting the apparatus so that the disturbance vanishes, this result carries the general absence of effect of the Earth's motion in optical experiments, up to the second order of small quantities.

Influence of translatory motion on the Structure of a Molecule: the law of Conservation of Mass

114. As a simple illustration of the general molecular theory, let us consider the group formed of a pair of electrons of opposite signs describing steady circular orbits round each other in a position of rest:[†] we can assert from the correlation, that when this pair is moving through the aether with velocity v in a direction lying in the plane of their orbits, these orbits relative to the translatory motion will be flattened along the direction of v to ellipticity $1-\frac{1}{2}v^2/c^2$, while there will be a first-order retardation of phase in each orbital motion when the electron is in front of the mean position combined with acceleration when behind it so that on the whole the period will be changed only in the second-order ratio $1+\frac{1}{2}v^2/c^2$. The specification of the orbital modification produced by the translatory motion, for the general case when the direction of that motion is inclined to the plane of the orbit, may be made similarly: it can also be extended to an ideal molecule constituted of any orbital system of electrons however complex. But this statement implies that the nucleus of the electron is merely a singular point in the aether, that there is nothing involved in it of the nature of inertia foreign to the aether: it also implies that there are no forces between the electrons other than those that exist through the mediation of the aether as here defined, that is other than electric forces.

The circumstance that the changes of their free periods, arising from convection of the molecules through the aether, are of the second order in v/c, is of course vital for the theory of the spectroscopic measurement of celestial velocities in the line of sight. That conclusion would however still hold good if we imagined the molecule to have inertia and potential energy extraneous to (*i.e.* unconnected with) the aether of optical and electrical phenomena,

[†] The orbital velocities are in this illustration supposed so small that radiation is not important. Cf. §§ 151-6 *infra*.

provided these properties are not affected by the uniform motion: for the aethereal fields of the moving electric charges, free or constrained, existing in the molecule, will be symmetrical fore and aft and unaltered to the first order by the motion, and therefore a change of sign of the velocity of translation will not affect them, so that the periods of free vibration cannot involve the first power of this velocity.

115. The fact that uniform motion of the molecule through the aether does not disturb its constitution to the first order, nor the aethereal symmetry of the moving system fore and aft, shows that when steady motion is established the mean kinetic energy of the system consists of the internal energy of the molecule, which is the same as when it is at rest, together with the sum of the energies belonging to the motions of translation of its separate electrons. This is verified on reflecting that the disturbance in the aether is made up additively of those due to the internal motions of the electrons in the molecule and those due to their common velocity of translation. Thus in estimating the mean value of the volume-integral of the square of the aethereal disturbance, which is the total kinetic energy, we shall have the integrated square of each of these disturbances separately, together with the integral of terms involving their product. Now one factor of this product is constant in time and symmetrical fore and aft as regards each electron, that factor namely which arises from the uniform translation; the other factor, arising from the orbital motions of the electrons, is oscillatory and symmetrical in front and rear of each orbit: thus the integrated product is by symmetry null. This establishes the result stated, that the kinetic energy of the moving molecule is made up of an internal energy, the same up to the first order of the ratio of its velocity to that of radiation as if it were at rest, and the energy of translation of its electrons. The coefficient of half the square of the velocity of translation in the latter part is therefore, up to that order, the measure of the inertia, or mass, of the molecule thus constituted. Hence when the square of the ratio

of the velocity of translation of the molecule to that of radiation is neglected, its electric inertia is equal to the sum of those of the electrons which compose it; and the fundamental chemical law of the constancy of mass throughout molecular transformations is verified for that part of the mass (whether it be all of it or not) that is of electric origin.

116. Objection has been taken to the view that the whole of the inertia of a molecule is associated with electric action, on the ground that gravitation, which has presumably no relations with such action, is proportional to mass: it has been suggested that inertia and gravity may be different results of the same cause. Now the inertia is by definition the coefficient of half the square of the velocity in the expression for the translatory energy of the molecule: in the constitution of the molecule it is admitted, from electrolytic considerations, that electric forces or agencies prevail enormously over gravitative ones: it seems fair to conclude that of its energy the electric part prevails equally over the gravitative part: but this is simply asserting that inertia is mainly of electric, or rather of aethereal, origin. Moreover the increase of kinetic electric energy of an electron arising from its motion with velocity v depends on v^2/c^2, on the coefficient of inertia of the aether, and on the dimensions of its nucleus, where c is the velocity of radiation: the increase of its gravitational energy would presumably in like manner depend on v^2/c'^2, where c' is the velocity of propagation of gravitation and is enormously greater than c. On neither ground does it appear likely that mass is to any considerable degree an attribute of gravitation.

.   .   .   .   .

10. AN INQUIRY INTO ELECTRICAL AND OPTICAL PHENOMENA IN MOVING BODIES*

H. A. LORENTZ

Introduction

§ 1. No answer satisfactory to all physicists has yet been found to the question whether or not the ether takes part in the movement of ponderable bodies.

To make a decision one must in the first instance rely on the aberration of light and the phenomena connected with it. However, so far neither of the two competing theories, either that of Fresnel or that of Stokes, have been completely successful with respect to all observations. Thus the choice between the two views must be made by weighing the remaining difficulties of one against those of the other. This is the way in which I came to the opinion quite some time ago that Fresnel's idea, hypothesizing a motionless ether, is on the right path. Of course, hardly more than one objection can be raised against the theory of Mr. Stokes, namely that his assumptions about the movement of the ether taking place in the neighbourhood of the earth contradict themselves.[†] How-

---

* Published by Brill, Leyden, 1895.
† Lorentz, De l'influence du mouvement de la terre sur les phénomènes lumineux. *Arch. néerl.* **21**, 103 (1887); Lodge, Aberration problems. *London Phil. Trans.* (A), **184**, 727 (1893); Lorentz, De aberratietheorie van Stokes. *Zittingsverslagen der Akad. v. Wet. te Amsterdam*, 97 (1892–3).

ever, that is a very weighty objection and I really cannot see how one can eliminate it.

Difficulties arise for the Fresnel theory in connection with the well known interference experiment of Mr. Michelson[†] and according to the opinion of some, also through the experiments with which Mr. Des Coudres has tried in vain to prove the influence of the movement of the earth on the mutual induction of two circuits.[‡] The results of the American scientist, however, can be explained with the aid of an auxiliary hypothesis and those found by Mr. Des Coudres can even be explained quite easily without such an auxiliary hypothesis.

As regards the observations of Mr. Fizeau[§] on the rotation of the plane of polarization in glass columns, that is another matter. At first glance, the result goes definitely against Stokes' theory. But when I attempted to develop Fresnel's theory further I also encountered difficulties in explaining Fizeau's experiments, and so I came gradually to suspect that the result of the experiments arose out of observational errors, or at least that the result did not correspond with the theoretical considerations on which the experiments were based. Mr. Fizeau has been kind enough to notify my colleague Mr. van de Sande-Bakhuizen, on the latter's inquiries, that at present he himself does not consider his observations to be decisive anymore.

In the course of this work I will return to the here mentioned questions in more detail. At the moment I am only interested in justifying for the time being the point of view which I have taken.

Several well known reasons can be cited in support of the Fresnel theory. First of all there is the impossibility of confining the ether between solid or liquid walls. As far as we know, a space contain-

---

[†] Michelson, *American Journal of Science* (3), **22**, 120; **34**, 333 (1887); *Phil. Mag.* (5), **24**, 449 (1887).
[‡] Des Coudres, *Wied. Ann.* **38**, 71 (1889).
[§] Fizeau, *Ann. de chim. et de phys.* (3), **58**, 129 (1860); *Pogg. Ann.* **114**, 554 (1861).

ing no air behaves, mechanically speaking, like a true vacuum where the movement of ponderable bodies is concerned. Seeing how the mercury in a barometer rises to the top of the tube when it is inclined or how easily a thin walled closed metal tube can be compressed, one cannot avoid the inference that solid and liquid bodies are completely permeable to the ether. One can hardly assume that this medium could undergo compression without offering resistance.

Fizeau's famous interference experiment with flowing water[†] proves that transparent substances can move without imparting their full velocity to the ether contained in them. It would have been impossible for this experiment—later repeated by Michelson and Morley on a larger scale[‡]—to have the observed result if *everything* present in one of the tubes were to have the same velocity. Only the behaviour of opaque substances and very extensive bodies remains problematic after this.

It is to be noted furthermore that the permeability of a body for the ether can be conceived in two ways. Firstly this property may be lacking in the single atom and yet appear in a larger mass if the atoms are extremely small compared with the spaces in between them. Secondly one can assume (and this is the hypothesis which I will use as a basis for my considerations) that ponderable matter is *absolutely* permeable, i.e., that the atom and the ether exist in the same place, which is conceivable if one could regard the atoms as local modifications of the ether.

It is not my intention to enter into such speculations in more detail or to make conjectures about the nature of the ether. My wish is only to keep myself as free as possible from pre-conceived opinions on this medium or to attribute to it, for example, any of the properties of ordinary liquids and gases. Should it turn out

---

[†] Fizeau, *Ann. de chim. et de phys.* (3), **57**, 385 (1859); *Pogg. Ann.* **3**, 457 (1853).

[‡] Michelson and Morley, *American Journal of Science* (3), **31**, 377 (1886).

that a description of the phenomena were to succeed best by assuming absolute permeability, then one ought to accept such an assumption for the present and leave it to future research to lead possibly to more profound understanding.

It is surely self-evident that there can be no question concerning the *absolute* rest of the ether. This expression would not even make sense. If I say for brevity's sake that the ether is at rest I only mean to say that one part of this medium does not move with respect to the other and that all observable motion of celestial bodies is motion relative to the ether.

§ 2. Since Maxwell's views have been increasingly accepted, the question about the behaviour of the ether has also become of great importance for the electromagnetic theory. In fact, strictly speaking, no experiment involving the motion of a charged body or an electrical conductor can be treated rigorously unless at the same time a statement is made about rest or motion of the ether. For an electric phenomenon the question arises whether an influence through the movement of the earth is to be expected, and, as regards the effect of the latter on optical phenomena, it is to be required of the electromagnetic theory of light that it takes into account all facts so far ascertained.

This is because the theory of aberration does not belong to those parts of Optics for the treatment of which the general principles of the wave theory are sufficient. As soon as a telescope is used one cannot get around applying Fresnel's convection coefficient to the lenses, and the value of this coefficient can only be derived from special assumptions concerning the nature of light waves.

The electromagnetic theory of light does in fact lead to the coefficient assumed by Fresnel as I demonstrated two years ago.[†] Since then I have simplified the theory considerably as well as extending it to phenomena of reflexion and refraction and also to

---

[†] Lorentz, *La théorie électromagnétique de Maxwell et son application aux corps mouvants*. Leiden, E. J. Brill, 1892.

double refracting substances.[†] I might therefore be allowed to return again to the subject.

In order to arrive at the fundamental equations for the electric phenomena in moving bodies I have taken a view which has been adopted by several physicists in the last years, i.e., I have assumed that all substances contain small electrically charged mass-particles and that all electric phenomena are based on the structure and movement of the "ions". This concept is, with respect to the electrolytes, generally recognized as the only possible one and Messrs. Giese,[‡] Schuster,[§] Arrhenius,[||] Elster and Geitel[††] have defended the view that also where the electric conductivity in gases is concerned one is dealing with a convection of ions. I do not see any obstacle to the assumption that the molecules of ponderable dielectric substances also contain such particles which are bound to definite equilibrium positions, and which can only be displaced from them through external electrical forces. This then would constitute the di-electric polarization of such substances.

The periodically alternating polarizations which according to Maxwell's theory constitute a ray of light, are attributed on this view to vibrations of the ions. It is well known that many scientists who accepted the view of the older theory of light considered the resonance of ponderable matter as the cause of dispersion, and this explanation can, in the main, be assimilated into the electromagnetic theory of light. In order to do this it is only necessary to ascribe a definite mass to the ions. I have shown this in a former treatise[‡‡] in which, it is true, I derived the fluctuation equation of

---

[†] Preliminary communications about this have appeared in the *Zittingsverlangen der Akad. v. Wet. te Amsterdam*, **28**, 149 (1892–3).

[‡] Giese, *Wied. Ann.* **17**, 538 (1882).

[§] Schuster, *Proc. Roy. Soc.* **37**, 317 (1884).

[||] Arrhenius, *Wied. Ann.* **32**, 565 (1887); **33**, 638 (1888).

[††] Elster and Geitel, *Wiener Sitz-Ber.* **97**, (2), 1255 (1888).

[‡‡] Lorentz, *Over het verband tusschen de voortplantingssnelheid van het licht on de dichtheid en samenstelling der middenstoffen. Verhandelingen der Akad. van Wet. te Amsterdam*, Deel 18, 1878; *Wied. Ann.* **9**, 641 (1880).

motion from action at a distance and not from Maxwellian concepts, which I now consider much simpler. Von Helmholtz[†] subsequently based his electromagnetic theory of dispersion on the same idea.[‡]

Mr. Giese[§] has applied to several cases the hypothesis that in metal conductors, too, the electricity is bound to ions. However, the picture which he draws of the phenomena in these substances differs in one point essentially from the concept which one has of conduction in electrolytes. While the particles of a dissolved salt—however frequently they may be arrested by the water molecules—can, after all, wander over long distances, the ions in the copper wire can hardly possess such great mobility. One can, however, be satisfied with a back and forth movement over molecular distances, provided one assumes that frequently one ion transmits its charge to another ion or that two ions of opposite charge exchange their charge when they meet or after they have "combined". In any case such processes must take place at the boundary between two substances when a current passes from one into the other. If for instance n positively charged copper atoms are deposited from a salt solution onto a copper plate and one considers also that in the copper plate all electricity is bound to ions, it is necessary to assume that the charges pass to n atoms in the plate or that $\frac{1}{2}n$ of the deposited particles have exchanged their charge with $\frac{1}{2}n$ negatively charged copper atoms already in the electrode.

If, therefore, the assumption of this transport or exchange of ion charges—admittedly still a highly unelucidated process—is an indispensible part of any theory which assumes the transport of electricity by ions, then a steady electric current never consists of convection alone, at least not when the centres of two touching or

[†] v. Helmholtz, *Wied. Ann.* **48**, 389 (1893).

[‡] Although in a different way, Mr. Koláček (*Wied. Ann.* **32**, 244, 429 [1887]) has also given an explanation of the dispersion of the electric waves in molecules. Also to be cited is the theory of Mr. Goldhammer (*Wied. Ann.* **7**, 93 [1892]).

[§] Giese, *Wied. Ann.* **37**, 576 (1889).

combined particles are at the distance l apart. The movement of electricity then takes place without convection over a distance of the order l, and only if this distance is very small in comparison to those over which convection takes place is one mainly concerned with the latter phenomenon.

Mr. Giese if of the opinion that a true convection does not take place at all in metals. As it seems, however, impossible to include the "jumping over" of charges in the theory, I beg to be excused if I myself completely avoid such a process and simply imagine a current in a copper wire as a movement of charged particles.

Further research will have to decide whether under a different assumption the results of the theory remain valid.

§ 3. The ionic theory was very appropriate for my purpose as it makes it possible to introduce the permeability for ether in a fairly satisfactory manner into the equations. These fall, of course, into two groups. First it is necessary to express how the state of the ether is determined by the charge, position and movement of the ions. Then, secondly, it must be noted with what forces the ether acts upon the charged particles. In my paper quoted above[†] I have derived the formulae from a few assumptions by means of D'Alembert's principle and thus chosen a way which is rather similar to Maxwell's use of Lagrange's equations. Now I prefer for brevity's sake to represent the fundamental equations themselves as hypotheses.

The formulae for the ether agree, as regards the space between the ions, with the well known equations of Maxwell's theory, and state in general that any change which is produced in the ether by an ion is propagated with the speed of light. However, we consider the force with which the ether acts on a charged particle to be dependent on the state of this medium at the position of the particle. The assumed fundamental law therefore differs in one essential point from the laws formulated by Weber and Clausius. The in-

[†] Lorentz, *La théorie électromagnétique de Maxwell et son application aux corps mouvants.*

fluence suffered by a particle B as a result of the proximity of a second particle A does depend on the movement of the latter, however not on its movement at that time. On the contrary the movement which this A had at a previous time is decisive, and the postulated law corresponds to the requirement which Gauss demanded in 1845 of the theory of electrodynamics which he expressed in his well known letter to Weber.[†]

Generally, we find, in the suppositions that I introduce, a return in a certain sense to the older theory of electricity. The kernel of Maxwell's views is not lost thereby, but it cannot be denied that one who subscribes to the ionic hypothesis is not very far removed from the notion of electrical particles which earlier theorists embraced. In certain simple cases this is particularly evident. Since we see the presence of electrical charge in an aggregate of positively or negatively charged particles, and our basic formulae yield Coulomb's Law for motionless ions, all, e.g., of electrostatics, can now be brought back to the earlier form.

[†] Gauss, *Werke*, **5**, 629.

11. SIMPLIFIED THEORY OF ELECTRICAL AND OPTICAL PHENOMENA IN MOVING SYSTEMS*

H. A. LORENTZ

§ 1. In former investigations I have assumed that, in all electrical and optical phenomena, taking place in ponderable matter, we have to do with small charged particles or ions, having determinate positions of equilibrium in dielectrics, but free to move in conductors except in so far as there is a resistance, depending on their velocities. According to these views an electric current in a conductor is to be considered as a progressive motion of the ions, and a dielectric polarization in a non-conductor as a displacement of the ions from their positions of equilibrium. The ions were supposed to be perfectly permeable to the aether, so that they can move while the aether remains at rest. I applied to the aether the ordinary electromagnetic equations, and to the ions certain other equations which seemed to present themselves rather naturally. In this way I arrived at a system of formulae which were found sufficient to account for a number of phenomena.

In the course of the investigation some artifices served to shorten the mathematical treatment. I shall now show that the theory may be still further simplified if the fundamental equations are immediately transformed in an appropriate manner.

§ 2. I shall start from the same hypotheses and introduce the

---

* *Proc. Roy. Acad. Amsterdam* **1,** 427 (1899).

same notations as in my "Versuch einer Theorie der electrischen und optischen Erscheinungen in bewegten Körpern". Thus, $\mathfrak{d}$ and $\mathfrak{H}$ will represent the dielectric displacement and the magnetic force, ϱ the density to which the ponderable matter is charged, $\mathfrak{v}$ the velocity of this matter, and $\mathfrak{E}$ the force acting on it per unit charge (electric force). It is only in the interior of the ions that the density ϱ differs from 0; for simplicity's sake I shall take it to be a continuous function of the coordinates, even at the surface of the ions. Finally, I suppose that each element of an ion retains its charge while it moves.

If, now, V be the velocity of light in the aether, the fundamental equations will be

$$Div\, \mathfrak{d} = \varrho, \tag{Ia}$$

$$Div\, \mathfrak{H} = 0, \tag{IIa}$$

$$Rot\, \mathfrak{H} = 4\pi\varrho\mathfrak{v} + 4\pi\dot{\mathfrak{d}}, \tag{IIIa}$$

$$4\pi V^2 Rot\, \mathfrak{d} = -\dot{\mathfrak{H}}, \tag{IVa}$$

$$\mathfrak{E} = 4\pi V^2 \mathfrak{d} + [\mathfrak{v}.\mathfrak{H}]. \tag{Va}$$

§ 3. We shall apply these equations to a system of bodies, having a common velocity of translation $\mathfrak{p}$, of constant direction and magnitude, the aether remaining at rest, and we shall henceforth denote by $\mathfrak{v}$, not the whole velocity of a material element, but the velocity it may have in addition to $\mathfrak{p}$

Now it is natural to use a system of axes of coordinates, which partakes of the translation $\mathfrak{p}$. If we give to the axis of x the direction of the translation, so that $\mathfrak{p}_y$ and $\mathfrak{p}_z$ are 0, the equations (Ia)–(Va) will have to be replaced by

$$Div\, \mathfrak{d} = \varrho, \tag{Ib}$$

$$Div\, \mathfrak{H} = 0, \tag{IIb}$$

$$\left.\begin{array}{l}\dfrac{\partial \mathfrak{H}_z}{\partial y}-\dfrac{\partial \mathfrak{H}_y}{\partial z} = 4\pi\varrho(\mathfrak{p}_x+\mathfrak{v}_x)+4\pi\left(\dfrac{\partial}{\partial t}-\mathfrak{p}_x\dfrac{\partial}{\partial x}\right)\mathfrak{d}_x, \\[4pt] \dfrac{\partial \mathfrak{H}_x}{\partial z}-\dfrac{\partial \mathfrak{H}_z}{\partial x} = 4\pi\varrho\mathfrak{v}_y+4\pi\left(\dfrac{\partial}{\partial t}-\mathfrak{p}_x\dfrac{\partial}{\partial x}\right)\mathfrak{d}_y, \\[4pt] \dfrac{\partial \mathfrak{H}_y}{\partial x}-\dfrac{\partial \mathfrak{H}_x}{\partial y} = 4\pi\varrho\mathfrak{v}_z+4\pi\left(\dfrac{\partial}{\partial t}-\mathfrak{p}_x\dfrac{\partial}{\partial x}\right)\mathfrak{d}_z,\end{array}\right\} \quad\text{(IIIb)}$$

$$\left.\begin{array}{l}4\pi V^2\left(\dfrac{\partial \mathfrak{d}_z}{\partial y}-\dfrac{\partial \mathfrak{d}_y}{\partial z}\right) = -\left(\dfrac{\partial}{\partial t}-\mathfrak{p}_x\dfrac{\partial}{\partial x}\right)\mathfrak{H}_x, \\[4pt] 4\pi V^2\left(\dfrac{\partial \mathfrak{d}_x}{\partial z}-\dfrac{\partial \mathfrak{d}_z}{\partial x}\right) = -\left(\dfrac{\partial}{\partial t}-\mathfrak{p}_x\dfrac{\partial}{\partial x}\right)\mathfrak{H}_y, \\[4pt] 4\pi V^2\left(\dfrac{\partial \mathfrak{d}_y}{\partial x}-\dfrac{\partial \mathfrak{d}_x}{\partial y}\right) = -\left(\dfrac{\partial}{\partial t}-\mathfrak{p}_x\dfrac{\partial}{\partial x}\right)\mathfrak{H}_z,\end{array}\right\} \quad\text{(IVb)}$$

$$\mathfrak{E} = 4\pi V^2\mathfrak{d}+[\mathfrak{p}\cdot\mathfrak{H}]+[\mathfrak{v}\cdot\mathfrak{H}]. \tag{Vb}$$

In these formulae the sign *Div*, applied to a vector $\mathfrak{A}$, has still the meaning defined by

$$Div\,\mathfrak{U} = \frac{\partial \mathfrak{A}_x}{\partial x}+\frac{\partial \mathfrak{A}_y}{\partial y}+\frac{\partial \mathfrak{A}_z}{\partial z}.$$

As has already been said, $\mathfrak{v}$ is the relative velocity with regard to the moving axes of coordinates. If $\mathfrak{v} = 0$, we shall speak of a system at rest; this expression therefore means relative rest with regard to the moving axes.

In most applications $\mathfrak{p}$ would be the velocity of the earth in its yearly motion.

§ 4. Now, in order to simplify the equations, the following quantities may be taken as independent variables

$$x' = \frac{V}{\sqrt{(V^2-\mathfrak{p}_x^2)}}\,x, \quad y' = y, \quad z' = z, \quad t' = t-\frac{\mathfrak{p}_x}{V^2-\mathfrak{p}_x^2}\,x. \tag{1}$$

The last of these is the time, reckoned from an instant that is not the same for all points of space, but depends on the place we wish to consider. We may call it the *local time*, to distinguish it from the *universal time t*.

If we put
$$\frac{V}{\sqrt{(V^2-\mathfrak{p}_x^2)}} = k,$$

we shall have

$$\frac{\partial}{\partial x} = k\frac{\partial}{\partial x'} - k^2\frac{\mathfrak{p}_x}{V^2}\frac{\partial}{\partial t'}, \quad \frac{\partial}{\partial y} = \frac{\partial}{\partial y'},$$

$$\frac{\partial}{\partial z} = \frac{\partial}{\partial z'}, \quad \frac{\partial}{\partial t} = \frac{\partial}{\partial t'}.$$

The expression
$$\frac{\partial \mathfrak{A}_x}{\partial x'} + \frac{\partial \mathfrak{A}_y}{\partial y'} + \frac{\partial \mathfrak{A}_z}{\partial z'}$$

will be denoted by
$$Div'\ \mathfrak{A}.$$

We shall also introduce, as new dependent variables instead of the components of $\mathfrak{d}$ and $\mathfrak{H}$, those of two other vectors $\mathfrak{F}'$ and $\mathfrak{H}$, which we define as follows

$$\mathfrak{F}'_x = 4\pi V^2 \mathfrak{d}_x, \quad \mathfrak{F}'_y = 4\pi k V^2 \mathfrak{d}_y - k\mathfrak{p}_x \mathfrak{H}_z, \quad \mathfrak{F}'_z = 4\pi k V^2 \mathfrak{d}_z + k\mathfrak{p}_x \mathfrak{H}_y,$$
$$\mathfrak{H}'_x = k\mathfrak{H}_x, \quad \mathfrak{H}'_y = k^2 \mathfrak{H}_y + 4\pi k^2 \mathfrak{p}_x \mathfrak{d}_z, \quad \mathfrak{H}'_z = k^2 \mathfrak{H}_z - 4\pi k^2 \mathfrak{p}_x \mathfrak{d}_y.$$

In this way I find by transformation and mutual combination of the equations (Ib)–(Vb):

$$Div'\ \mathfrak{F}' = \frac{4\pi}{k} V^2 \varrho - 4\pi k \mathfrak{p}_x \varrho \mathfrak{v}_x, \tag{Ic}$$

$$Div'\ \mathfrak{H}' = 0, \tag{IIc}$$

$$\left.\begin{aligned}
\frac{\partial \mathfrak{H}'_z}{\partial y'} - \frac{\partial \mathfrak{H}'_y}{\partial z'} &= 4\pi k^2 \varrho \mathfrak{v}_x + \frac{k^2}{V^2}\frac{\partial \mathfrak{F}'_x}{\partial t'} \\
\frac{\partial \mathfrak{H}'_x}{\partial z'} - \frac{\partial \mathfrak{H}'_z}{\partial x'} &= 4\pi k \varrho \mathfrak{v}_y + \frac{k^2}{V^2}\frac{\partial \mathfrak{F}'_y}{\partial t'} \\
\frac{\partial \mathfrak{H}'_y}{\partial x'} - \frac{\partial \mathfrak{H}'_x}{\partial y'} &= 4\pi k \varrho \mathfrak{v}_z + \frac{k^2}{V^2}\frac{\partial \mathfrak{F}'_z}{\partial t'}
\end{aligned}\right\}, \tag{IIIc}$$

$$\left.\begin{array}{l}\dfrac{\partial \mathfrak{F}'_z}{\partial y'} - \dfrac{\partial \mathfrak{F}'_y}{\partial z'} = -\dfrac{\partial \mathfrak{H}'_x}{\partial t'} \\[4pt] \dfrac{\partial \mathfrak{F}'_x}{\partial z'} - \dfrac{\partial \mathfrak{F}'_z}{\partial x'} = -\dfrac{\partial \mathfrak{H}'_y}{\partial t'} \\[4pt] \dfrac{\partial \mathfrak{F}'_y}{\partial x'} - \dfrac{\partial \mathfrak{F}'_x}{\partial y'} = -\dfrac{\partial \mathfrak{H}'_z}{\partial t'}\end{array}\right\}, \qquad \text{(IVc)}$$

$$\left.\begin{array}{l}\mathfrak{E}_x = \mathfrak{F}'_x + k\,\dfrac{\mathfrak{p}_x}{V^2}(\mathfrak{v}_y\mathfrak{F}'_y + \mathfrak{v}_z\mathfrak{F}'_z) + (\mathfrak{v}_y\mathfrak{H}'_z - \mathfrak{v}_z\mathfrak{H}'_y) \\[4pt] \mathfrak{E}_y = \dfrac{1}{k}\mathfrak{F}'_y - k\,\dfrac{\mathfrak{p}_x}{V^2}\mathfrak{v}_x\mathfrak{F}'_y + \left(\dfrac{1}{k}\mathfrak{v}_z\mathfrak{H}'_x - \mathfrak{v}_x\mathfrak{H}'_z\right) \\[4pt] \mathfrak{E}_z = \dfrac{1}{k}\mathfrak{F}'_z - k\,\dfrac{\mathfrak{p}_x}{V^2}\mathfrak{v}_x\mathfrak{F}'_z + \left(\mathfrak{v}_x\mathfrak{H}'_y - \dfrac{1}{k}\mathfrak{v}_y\mathfrak{H}'_x\right)\end{array}\right\}. \qquad \text{(Vc)}$$

Putting $\mathfrak{v} = 0$ in the three last equations we see that

$$\mathfrak{F}'_x, \quad \frac{1}{k}\mathfrak{F}'_y, \quad \frac{1}{k}\mathfrak{F}'_z$$

are the components of the electric force that would act on a particle at rest.

§ 5. We shall begin with an application of the equations to electrostatic phenomena. In these we have $\mathfrak{v} = 0$ and $\mathfrak{F}'$ independent of the time. Hence, by (IIc) and (IIIc)

$$\mathfrak{H}' = 0,$$

and by (IVc) and (Ic)

$$\frac{\partial \mathfrak{F}'_z}{\partial y'} - \frac{\partial \mathfrak{F}'_y}{\partial z'} = 0, \quad \frac{\partial \mathfrak{F}'_x}{\partial z'} - \frac{\partial \mathfrak{F}'_z}{\partial x'} = 0, \quad \frac{\partial \mathfrak{F}'_y}{\partial x'} - \frac{\partial \mathfrak{F}'_x}{\partial y'} = 0,$$

$$\text{Div}'\,\mathfrak{F}' = \frac{4\pi}{k}V^2\varrho.$$

These equations show that $\mathfrak{F}'$ depends on a potential ω, so that

$$\mathfrak{F}'_x = -\frac{\partial \omega}{\partial x'}, \quad \mathfrak{F}'_y = -\frac{\partial \omega}{\partial y'}, \quad \mathfrak{F}'_z = -\frac{\partial \omega}{\partial z'}$$

and
$$\frac{\partial^2 \omega}{\partial x'^2} + \frac{\partial^2 \omega}{\partial y'^2} + \frac{\partial^2 \omega}{\partial z'^2} = -\frac{4\pi}{k} V^2 \varrho. \tag{2}$$

Let S be the system of ions with the translation $\mathfrak{p}_x$, to which the above formulae are applied. We can conceive a second system S_0 with no translation and consequently no motion at all; we shall suppose that S is changed into S_0 by a dilatation in which the dimensions parallel to OX are changed in ratio of 1 to k, the dimensions perpendicular to OX remaining what they were. Moreover we shall attribute equal charges to corresponding volume-elements in S and S_0; if then ϱ_0 be the density in a point P of S, the density in the corresponding point P_0 of S_0 will be

$$\varrho_0 = \frac{1}{k} \varrho.$$

If x, y, z are the coordinates of P, the quantities x', y', z', determined by (1), may be considered as the coordinates of P_0.

In the system S_0, the electric force, which we shall call $\mathfrak{E}_0$ may evidently be derived from a potential ω_0, by means of the equations

$$\mathfrak{E}_{0x} = -\frac{\partial \omega_0}{\partial x'}, \quad \mathfrak{E}_{0y} = -\frac{\partial \omega_0}{\partial y'}, \quad \mathfrak{E}_{0z} = -\frac{\partial \omega_0}{\partial z'},$$

and the function ω_0 itself will satisfy the condition

$$\frac{\partial^2 \omega_0}{\partial x'^2} + \frac{\partial^2 \omega_0}{\partial y'^2} + \frac{\partial^2 \omega_0}{\partial z'^2} = -4\pi V^2 \varrho_0 = -\frac{4\pi}{k} V^2 \varrho.$$

Comparing this with (2), we see that in corresponding points

$$\omega = \omega_0,$$

and consequently

$$\mathfrak{F}'_x = \mathfrak{E}_{0x}, \quad \mathfrak{F}'_y = \mathfrak{E}_{0y}, \quad \mathfrak{F}'_z = \mathfrak{E}_{0z}.$$

In virtue of what has been remarked at the end of § 4, the components of the electric force in the system S will therefore be

$$\mathfrak{E}_{0x}, \quad \frac{1}{k}\mathfrak{E}_{0y}, \quad \frac{1}{k}\mathfrak{E}_{0z}.$$

Parallel to OX we have the same electric force in S and S_0, but in a direction perpendicular to OX the electric force in S will be $1/k$ times the electric force in S_0.

By means of this result every electrostatic problem for a moving system may be reduced to a similar problem for a system at rest; only the dimensions in the direction of translation must be slightly different in the two systems. If, e.g., we wish to determine in what way innumerable ions will distribute themselves over a moving conductor C, we have to solve the same problem for a conductor C_0, having no translation. It is easy to show that if the dimensions of C_0 and C differ from each other in the way that has been indicated, the electric force in one case will be perpendicular to the surface of C, as soon as, in the other case, the force $\mathfrak{E}_0$ is normal to the surface of C_0.

Since

$$k = \left(1 - \frac{\mathfrak{p}_x^2}{V^2}\right)^{-\frac{1}{2}}$$

exceeds unity only by a quantity of the second order—if we call $\mathfrak{p}_x/V$ of the first order—the influence of the Earth's yearly motion on electrostatic phenomena will likewise be of the second order.

§ 6. We shall now shew how our general equations (Ic)—(Vc) may be applied to optical phenomena. For this purpose we consider a system of ponderable bodies, the ions in which are capable of vibrating about determinate positions of equilibrium. If the system be traversed by waves of light, there will be oscillations of the ions, accompanied by electric vibrations in the aether. For convenience of treatment we shall suppose that, in the absence of light-waves, there is no motion at all; this amounts to ignoring all molecular motion.

Our first step will be to omit all terms of the second order. Thus, we shall put $k = 1$, and the electric force acting on ions at rest will become $\mathfrak{F}'$ itself.

We shall further introduce certain restrictions, by means of which we get rid of the last term in (Ic) and of the terms containing $\mathfrak{v}_x, \mathfrak{v}_y, \mathfrak{v}_z$ in (Vc).

The first of these restrictions relates to the magnitude of the displacements $\mathfrak{a}$ from the positions of equilibrium. We shall suppose them to be exceedingly small, even relatively to the dimensions of the ions and we shall on this ground neglect all quantities which are of the second order with respect to $\mathfrak{a}$.

It is easily seen that, in consequence of the displacements, the electric density in a fixed point will no longer have its original value ϱ_0, but will have become

$$\varrho = \varrho_0 - \frac{\partial}{\partial x}(\varrho_0 \mathfrak{a}_x) - \frac{\partial}{\partial y}(\varrho_0 \mathfrak{a}_y) - \frac{\partial}{\partial z}(\varrho_0 \mathfrak{a}_z).$$

Here, the last terms, which evidently must be taken into account, have the order of magnitude $c\varrho_0/a$, if c denotes the amplitude of the vibrations; consequently, the first term of the right-hand member of (Ic) will contain quantities of the order

$$\frac{V^2 c \varrho_0}{a}. \tag{3}$$

On the other hand, if T is the time of vibration, the last term in (Ic) will be of the order

$$\frac{\mathfrak{p}_x \varrho_0 c}{T}. \tag{4}$$

Dividing this by (3), we get

$$\frac{\mathfrak{p}_x}{V} \cdot \frac{a}{VT},$$

an extremely small quantity, because the diameter of the ions is a very small fraction of the wave-length. This is the reason why we may omit the last term in (Ic).

As to the equations (Vc), it must be remarked that, if the displacements are infinitely small, the same will be true of the velocities and, in general, of all quantities which do not exist as long as the system is at rest and are entirely produced by the motion. Such are $\mathfrak{H}'_x$, $\mathfrak{H}'_y$, $\mathfrak{H}'_z$. We may therefore omit the last terms in (Vc), as being of the second order.

The same reasoning would apply to the terms containing $\mathfrak{p}_x/V^2$, if we could be sure that in the state of equilibrium there are no electric forces at all. If, however, in the absence of any vibrations, the vector $\mathfrak{F}'$ has already a certain value $\mathfrak{F}'_0$, it will only be the difference $\mathfrak{F}' - \mathfrak{F}'_0$, that may be called infinitely small; it will then be permitted to replace $\mathfrak{F}'_y$ and $\mathfrak{F}'_z$ by $\mathfrak{F}'_{0y}$ and $\mathfrak{F}'_{0z}$.

Another restriction consists in supposing that an ion is incapable of any motion but a translation as a whole, and that, in the position of equilibrium, though its parts may be acted on by electric forces, as has just been said, yet the whole ion does not experience a resultant electric force. Then, if $d\tau$ is an element of volume, and the integrations are extended all over the ion,

$$\int \varrho_0 \mathfrak{F}'_{0y}\, d\tau = \int \varrho_0 \mathfrak{F}'_{0z}\, d\tau = 0. \tag{5}$$

Again, in the case of vibrations, the equations (Vc) will only serve to calculate the resultant force acting on an ion. In the direction of the axis of y e.g. this force will be

$$\int \varrho \mathfrak{F}_y\, d\tau.$$

Its value may be found, if we begin by applying the second of the three equations to each point of the ion, always for the same universal time t, and then integrate. From the second term on the right-hand side we find

$$-\frac{\mathfrak{p}_x}{V^2}\int \varrho\mathfrak{v}_x\mathfrak{F}'_y\,d\tau,$$

or, since we may replace $\mathfrak{F}'_y$ by $\mathfrak{F}'_{0y}$ and ϱ by ϱ_0,

$$-\frac{\mathfrak{p}_x}{V^2}\mathfrak{v}_x\int \varrho_0\mathfrak{F}'_{0y}\,d\tau,$$

which vanishes on account of (5).

Hence, as far as regards the resultant force, we may put $\mathfrak{E} = \mathfrak{F}'$, that is to say, we may take $\mathfrak{F}'$ as the electric force, acting not only on ions at rest, but also on moving ions.

The equations will be somewhat simplified, if, instead of $\mathfrak{F}'$, we introduce the already mentioned difference $\mathfrak{F}' - \mathfrak{F}'_0$. In order to do this, we have only twice to write down the equations (Ic)–(IVc), once for the vibrating system and a second time for the same system in a state of rest; and then to subtract the equations of the second system from those of the first. In the resulting equations, I shall, for the sake of brevity, write $\mathfrak{F}'$ instead of $\mathfrak{F}' - \mathfrak{F}'_0$, so that henceforth $\mathfrak{F}'$ will denote not the total electric force, but only the part of it that is due to the vibrations. At the same time we shall replace the value of ϱ, given above, by

$$\varrho_0 - \mathfrak{a}_x\frac{\partial\varrho_0}{\partial x'} - \mathfrak{a}_y\frac{\partial\varrho_0}{\partial y'} - \mathfrak{a}_z\frac{\partial\varrho_0}{\partial z'}.$$

We may do so, because we have supposed $\mathfrak{a}_x, \mathfrak{a}_y, \mathfrak{a}_z$ to have the same values all over an ion, and because ϱ_0 is independent of the time, so that

$$\frac{\partial\varrho_0}{\partial x} = \frac{\partial\varrho_0}{\partial x'}.$$

Finally we have

$$Div'\,\mathfrak{F}' = -4\pi V^2\left(\mathfrak{a}_x\frac{\partial\varrho_0}{\partial x'} + \mathfrak{a}_y\frac{\partial\varrho_0}{\partial y'} + \mathfrak{a}_z\frac{\partial\varrho_0}{\partial z'}\right), \quad \text{(Id)}$$

$$Div' \, \mathfrak{H}' = 0, \tag{IId}$$

$$\frac{\partial \mathfrak{H}'_z}{\partial y'} - \frac{\partial \mathfrak{H}'_y}{\partial z'} = 4\pi\varrho_0 \, \frac{\partial \mathfrak{a}_x}{\partial t'} + \frac{1}{V^2} \, \frac{\partial \mathfrak{F}'_x}{\partial t'}, \quad \text{etc.} \tag{IIId}$$

$$\frac{\partial \mathfrak{F}'_z}{\partial y'} - \frac{\partial \mathfrak{F}'_y}{\partial z'} = -\frac{\partial \mathfrak{H}'_x}{\partial t'}, \quad \text{etc.} \tag{IVd}$$

Since these equations do no longer explicitly contain the velocity $\mathfrak{p}_x$, they will hold, without any change of form, for a system that has no translation, in which case, of course, t' would be the same thing as the universal time t.

Yet, strictly speaking, there would be a slight difference in the formulae, when applied to the two cases. In the system without a translation $\mathfrak{a}_x, \mathfrak{a}_y^\tau, \mathfrak{a}_z$ would be, in all points of an ion, the same functions of t', i.e. of the universal time, whereas, in the moving system, these components would not depend in the same way on t' in different parts of the ion, just because they must everywhere be the same functions of t.

However, we may ignore this difference, of the ions are so small, that we may assign to each of them a single local time, applicable to all its parts.

The equality of form of the electromagnetic equations for the two cases of which we have spoken will serve to simplify to a large extent our investigation. However, it should be kept in mind, that, to the equations (Id)–(IVd) we must add the equations of motion for the ions themselves. In establishing these, we have to take into account, not only the electric forces, but also all other forces acting on the ions. We shall call these latter the molecular forces and we shall begin by supposing them to be sensible only at such small distances, that two particles of matter, acting on each other, may be said to have the same local time.

§ 7. Let us now imagine two systems of ponderable bodies, the one S with a translation, and the other one S_0 without such a motion, but equal to each other in all other respects. Since we neglect quantities of the order $\mathfrak{p}_x^2/V^2$, the electric force will, by § 5 be the same in both systems, as long as there are no vibrations.

After these have been excited, we shall have for both systems the equations (Id)–(IVd).

Further we shall imagine motions of such a kind, that, if in a point (x', y', z') of S_0 we find a certain quantity of matter or a certain electric charge at the universal time t', an equal quantity of matter or an equal charge will be found in the corresponding point of S at the local time t'. Of course, this involves that at these corresponding times we shall have, in the point (x', y', z') of both systems, the same electric density, the same displacement $\mathfrak{a}$, and equal velocities and accelerations.

Thus, some of the dependent variables in our equations (Id)–(IVd) will be represented in S_0 and S by the same functions of x', y', z', t', whence we conclude that the equations will be satisfied by values of $\mathfrak{H}'_x, \mathfrak{H}'_y, \mathfrak{H}'_z, \mathfrak{F}'_x, \mathfrak{F}'_y, \mathfrak{F}'_z$, which are likewise in both cases the same functions of x', y', z', t'. By what has been said at the beginning of this §, not only $\mathfrak{F}'$, but also the total electric force will be the same in S_0 and S, always provided that corresponding ions at corresponding times (i.e. for equal values of t') be considered.

As to the molecular forces, acting on an ion, they are confined to a certain small space surrounding it, and by what has been said in § 6, the difference of local times within this space may be neglected. Moreover, if equal spaces of this kind are considered in S_0 and S, there will be, at corresponding times, in both the same distribution of matter. This is a consequence of what has been supposed concerning the two motions.

Now, the simplest assumption we can make on the molecular forces is this, that they are *not* changed by the translation of the system. If this be admitted, it appears from the above considerations that corresponding ions in S_0 and S will be acted on by the same molecular forces, as well as by the same electric forces. Therefore, since the masses and accelerations are the same, the supposed motion in S will be possible as soon as the corresponding motion in S_0 can really exist. In this way we are led to the following theorem.

If, in a body or a system of bodies, without a translation, a system of vibrations be given, in which the displacements of the ions and the components of $\mathfrak{F}'$ and $\mathfrak{H}'$ are certain functions of the coordinates and the time, then, if a translation be given to the system, there can exist vibrations, in which the displacements and the components of $\mathfrak{F}'$ and $\mathfrak{H}'$ are the same functions of the coordinates and the *local* time. This is the theorem, to which I have been led in a much more troublesome way in my "Versuch einer Theorie, etc.", and by which most of the phenomena, belonging to the theory of aberration may be explained.

§ 8. In what precedes, the molecular forces have been supposed to be confined to excessively small distances. If two particles of matter were to act upon each other at such a distance that the difference of their local times might not be neglected, the theorem would no longer be true in the case of molecular forces that are not altered at all by the translation. However, one soon perceives that the theorem would again hold good, if these forces were changed by the translation in a definite way, in such a way namely that the action between two quantities of matter were determined, not by the *simultaneous* values of their coordinates, but by their values at *equal local times*. If therefore, we should meet with phenomena, in which the difference of the local times for mutually acting particles might have a sensible influence, and in which yet observation showed the above theorem to be true, this would indicate a modification, like the one we have just specified, of the molecular forces by the influence of a translation. Of course, such a modification would only be possible, if the molecular forces were no direct actions at a distance, but were propagated by the aether in a similar way as the electromagnetic actions. Perhaps the rotation of the plane of polarization in the so-called active bodies will be found to be a phenomenon of the kind just mentioned.

§ 9. Hitherto all quantities of the order $\mathfrak{p}_x^2/V^2$ have been neglected.

As is well known, these must be taken into account in the discussion of MICHELSON's experiment, in which two rays of light interfered after having traversed rather long paths, the one parallel to the direction of the earth's motion, and the other perpendicular to it. In order to explain the negative result of this experiment FITZGERALD and myself have supposed that, in consequence of the translation, the dimensions of the solid bodies serving to support the optical apparatus, are altered in a certain ratio.

Some time ago, M. LIÉNARD[†] has emitted the opinion that, according to my theory, the experiment should have a positive result, if it were modified in so far that the rays had to pass through a solid or a liquid dielectric.

It is impossible to say with certainty what would be observed in such a case, for, if the explication of MICHELSON's result which I have proposed is accepted, we must also assume that the mutual distances of the molecules of transparent media are altered by the translation.

Besides, we must keep in view the possibility of an influence, be it of the second order, of the translation on the molecular forces.

In what follows I shall shew, not that the result of the experiment must necessarily be negative, but that this might very well be the case. At the same time it will appear what would be the theoretical meaning of such a result.

Let us return again to the equations (Ic)–(Vc). This time we shall not put in them $k = 1$, but the other simplifications of which we have spoken in § 6 will again be introduced. We shall now have to distinguish between the vectors $\mathfrak{E}$ and $\mathfrak{F}'$, the former alone being the electric force. By both signs I shall now denote, not the whole vector, but the part that is due to the vibrations.

The equations may again be written in a form in which the velocity of translation does not explicitly appear. For this purpose, it is necessary to replace the variables x', y', z', t', $\mathfrak{F}'$, $\mathfrak{H}'$, a and ϱ_0

[†] L'Éclairage Électrique, 20 et 27 août 1898.

by new ones, differing from the original quantities by certain constant factors.

For the sake of uniformity of notation all these new variables will be distinguished by double accents. Let ε be an indeterminate coefficient, differing from unity by a quantity of the order $\mathfrak{p}_x^2/V^2$, and let us put

$$x = \frac{\varepsilon}{k} x'', \quad y = \varepsilon y'', \quad z = \varepsilon z'', \tag{6}$$

$$\mathfrak{a}_x = \frac{\varepsilon}{k} \mathfrak{a}_x'', \quad \mathfrak{a}_y = \varepsilon \mathfrak{a}_y'', \quad \mathfrak{a}_z = \varepsilon \mathfrak{a}_z'', \tag{7}$$

$$\varrho_0 = \frac{k}{\varepsilon^3} \varrho_0'', \tag{8}$$

$$\mathfrak{F}_x' = \frac{1}{\varepsilon^2} \mathfrak{F}_x'', \quad \mathfrak{F}_y' = \frac{1}{\varepsilon^2} \mathfrak{F}_y'', \quad \mathfrak{F}_z' = \frac{1}{\varepsilon^2} \mathfrak{F}_z'',$$

$$\mathfrak{H}_x = \frac{k}{\varepsilon^2} \mathfrak{H}_x'', \quad \mathfrak{H}_y' = \frac{k}{\varepsilon^2} \mathfrak{H}_y'', \quad \mathfrak{H}_z' = \frac{k}{\varepsilon^2} \mathfrak{H}_z'',$$

$$t' = k\varepsilon t'', \tag{9}$$

so that t'' is a modified local time; then we find

$$Div'' \mathfrak{F}'' = 4\pi V^2 \left(-\mathfrak{a}_x'' \frac{\partial \varrho_0''}{\partial x''} - \mathfrak{a}_y'' \frac{\partial \varrho_0''}{\partial y''} - \mathfrak{a}_z'' \frac{\partial \varrho_0''}{\partial z''} \right), \tag{Ie}$$

$$Div'' \mathfrak{H}'' = 0, \tag{IIe}$$

$$\frac{\partial \mathfrak{H}_z''}{\partial y''} - \frac{\partial \mathfrak{H}_y''}{\partial z''} = 4\pi \varrho_0'' \frac{\partial \mathfrak{a}_x''}{\partial t''} + \frac{1}{V^2} \frac{\partial \mathfrak{F}_x''}{\partial t''}, \quad \text{etc.} \tag{IIIe}$$

$$\frac{\partial \mathfrak{F}_z''}{\partial y''} - \frac{\partial \mathfrak{F}_y''}{\partial z''} = -\frac{\partial \mathfrak{H}_x''}{\partial t''}, \quad \text{etc.} \tag{IVe}$$

$$\mathfrak{E}_x = \frac{1}{\varepsilon^2} \mathfrak{F}_x'', \quad \mathfrak{E}_y = \frac{1}{k\varepsilon^2} \mathfrak{F}_y'', \quad \mathfrak{E}_z = \frac{1}{k\varepsilon^2} \mathfrak{F}_z''. \tag{Ve}$$

These formulae will also hold for a system without translation; only, in this case we must take $k = 1$, and we shall likewise take

$\varepsilon = 1$, though this is not necessary. Thus, x'', y'', z'' will then be the coordinates, t'' the same thing as t, i.e. the universal time, $\mathfrak{a}''$ the displacement, ϱ_0'' the electric density, $\mathfrak{H}''$ and ψ'' the magnetic and electric forces, the last in so far as it is due to the vibrations.

Our next object will be to ascertain under what conditions, now that we retain the terms with $\mathfrak{p}_x^2/V^2$, two systems S and S_0, the first having a translation, and the second having none, may be in vibratory states that are related to each other in some definite way. This investigation resembles much the one that has been given in § 7; it may therefore be expressed in somewhat shorter terms.

To begin with, we shall agree upon the degree of similarity there shall be between the two systems in their states of equilibrium. In this respect we define S by saying that the system S_0 may be changed into it by means of the dilatations indicated by (6); we shall suppose that, in undergoing these dilatations, each element of volume retains its ponderable matter, as well as its charge. It is easily seen that this agrees with the relation (8).

We shall not only suppose that the system S_0 *may* be changed in this way into an imaginary system S, but that, as soon as the translation is given to it, the transformation *really* takes place, of itself, i.e. by the action of the forces acting between the particles of the system, and the aether. Thus, after all, S will be the *same* material system as S.

The transformation of which I have now spoken, is precisely such a one as is required in my explication of MICHELSON's experiment. In this explication the factor ε may be left indeterminate. We need hardly remark that for the real transformation produced by a translatory motion, the factor should have a definite value. I see, however, no means to determine it.

Before we proceed further, a word on the electric forces in S and S_0 in their states of equilibrium. If $\varepsilon = 1$, the relation between these forces will be given by the equations of § 5. Now ε indicates an alteration of all dimensions in the same ratio, and it is very easy to see what influence this will have on the electric forces.

Thus, it will be found that, in passing from S_0 to S, the electric force in the direction of OX will be changed in the ratio of 1 to $1/\varepsilon^2$, and that the corresponding ratio for the other components will be as 1 to $1/k\varepsilon^2$.

As to the corresponding vibratory motions, we shall require that at corresponding times, i.e. for equal values of t'', the configuration of S may always be got from that of S_0 by the above mentioned dilatations. Then, it appears from (7) that $\mathfrak{a}''_x$, $\mathfrak{a}''_y$, $\mathfrak{a}''_z$ will be, in both systems, the same functions of x'', y'', z'', t'', whence we conclude that the equations (Ie)–(IVe) can be satisfied by values of $\mathfrak{F}''_x$, $\mathfrak{H}''_x$, etc., which are likewise, in S_0 and in S, the same functions of x'', y'', z'', t''.

Always provided that we start from a vibratory motion in S_0 that can really exist, we have now arrived at a motion in S, that is possible in so far as it satisfies the electromagnetic equations. The last stage of our reasoning will be to attend to the molecular forces. In S_0 we imagine again, around one of the ions, the same small space we have considered in § 7, and to which the molecular forces acting on the ion are confined; in the other system we shall now conceive the corresponding small space, i.e. the space that may be derived from the first one by applying to it the dilatations (6). As before, we shall suppose these spaces to be so small that in the second of them there is no necessity to distinguish the local times in its different parts; then we may say that in the two spaces there will be, at corresponding times, corresponding distributions of matter.

We have already seen that, in the states of equilibrium, the electric forces parallel to OX, OY, OZ, existing in S differ from the corresponding forces in S_0 by the factors

$$\frac{1}{\varepsilon^2}, \quad \frac{1}{k\varepsilon^2} \quad \text{and} \quad \frac{1}{k\varepsilon^2}.$$

From (Ve) it appears that the same factors come into play when we consider the part of the electric forces that is due to the vibrations. If, now, we suppose that the molecular forces are modified in quite the same way in consequence of the translation, we may apply the just mentioned factors to the components of the *total* force acting on an ion. Then, the imagined motion in S will be a possible one, provided that these same factors to which we have been led in examining the forces present themselves again, when we treat of the product of the masses and the accelerations.

According to our suppositions, the accelerations in the directions of OX, OY, OZ in S are resp. $1/k^3\varepsilon$, $1/k^2\varepsilon$ and $1/k^2\varepsilon$ times what they are in S_0. If therefore the required agreement is to exist with regard to the vibrations parallel to OX, the ratio of the masses of the ions in S and S_0 should be k^3/ε; on the contrary we find for this ratio k/ε, if we consider in the same way the forces and the accelerations in the directions of OY and OZ.

Since k is different from unity, these values cannot both be 1; consequently, states of motion, related to each other in the way we have indicated, will only be possible if in the transformation of S_0 into S the masses of the ions change; even this must take place in such a way that the same ion will have different masses for vibrations parallel and perpendicular to the velocity of translation.

Such a hypothesis seems very startling at first sight. Nevertheless we need not wholly reject it. Indeed, as is well known, the *effective* mass of an ion depends on what goes on in the aether; it may therefore very well be altered by a translation and even to different degrees for vibrations of different directions.

If the hypothesis might be taken for granted, MICHELSON's experiment should always give a negative result, whatever transparent media were placed on the path of the rays of light, and even if one of these went through air, and the other, say through glass. This is seen by remarking that the correspondence between the two motions we have examined is such that, if in S_0 we had a certain distribution of light and dark (interference-bands) we should have in S a similar distribution, which might be got from that in S_0 by the dilatations (6), provided however that in S the time of vibration be $k\varepsilon$ times as great as in S_0. The necessity of this last difference follows from (9). Now the number $k\varepsilon$ would be the same in all positions we can give to the apparatus; therefore, if we continue to use the *same* sort of light, while rotating the instruments, the interference-bands will never leave the parts of the ponderable system, e.g. the lines of a micrometer, with which they coincided at first.

We shall conclude by remarking that the alteration of the molecular forces that has been spoken of in this § would be one of the second order, so that we have not come into contradiction with what has been said in § 7.

INDEX

Aberration Chapter III, 91, 100, 105, Papers 1, 2, 3
Aberration constant 22, 32
Abraham, M. 95
Action-at-a-distance 10, 76–7, 80
Ad hoc
   elimination of longitudinal wave in aether 17
   restrictions on radiating atoms 96
Aether wind velocity 30, 35
Airy, G. B. 176–7, 185–6
Airy's water-filled telescope experiment 26
Ampère, A. M. 96
Analogy 74–5, 89, 107, 116 *see also* Mechanical model(s)
Arago, F. 14, 23–5, 125

Basset, A. B. 67
Biaxal crystals 13, 15, 17
Biot, J. B. 13
Bjerknes, C. viii
Black-body radiation 98
Bork, A. 85n, 104n
Boscovich, R. G. 201
   view of atoms adopted by Faraday 77–8
   water-filled telescope (or microscope) 26, 132–4
Boussinesq, J. 67–8
Bradley's corpuscular explanation of aberration 22–3
Bradley's discovery of stellar aberration 19, 21–2
Brewster, D. 13
Bromberg, J. viii, 79
Brush, S. 104n

Central forces, principle of
   criticized by Green as too restrictive 46, 161
Challis, J. 141, 143
Classical mechanics 99 *see also* Mechanical explanation, Lagrangian mechanics, and Hamilton's Principle of Least Action
Clausius, R. 253
Corpuscular theory of light (emission theory) 9, 13, 22–4, 125, 136, 144–5, 213
Cotes' preface to Newton's *Principia* 10–11

D'Alembert's Principle of Virtual Velocities 41–2, 46–8, 162, 164, 253
Descartes, R. 7, 8, 10
Diffraction (inflection)
   Young 12
   Fresnel's theory 14–15
Dirac, P. A. M. 116–7

Dispersion 20, 100, 251
  ignored by Green 164
Doppler effect 20, 226, 241, 243
Double refraction 9, 12, 13, *15–17*, 58, 71, 98

Ehrenfest, P. 82
Einstein, A.
  special theory of relativity vii, viii, 4–5, *114–6*
Electromagnetic aether 45, 66, *76–98*, Papers 7, 9
Electromagnetic "Weltbild" 93, 95, 228
Electron (ion)
  Larmor's characterization as an aether singularity 95–6, 217–8, 224–5
  Larmor's incorporation into the MacCullagh–Fitzgerald aether 96–8
  Lorentz' view of 101–2, 256
  See also Mass
Electronic state 77–8
Energy
  of Maxwell's aether (electromagnetic field) 81, 83–5
  potential function in aether
    of Green 52, 164–9
    of MacCullagh 62–3, 190–93
Euler, L. 213

Faraday, M. 4, *77–8*, 227
Fermat's Principle 213
Feyerabend, P. K. 5–6
Fitzgerald, G. F. vii, 45, 66, *84*, 95
Fitzgerald's electromagnetic aether *84–9*, 90, Paper 7
Fizeau's experiment on rotation of plane of polarization in glass columns (pack of glass plates) 103, 248
Fizeau's moving water experiment 29, 145, 249

Fresnel, A. vii, 55, 174, 177, 188, 193
  aberration theory *23–9*, 145, 155, 247–8, Paper 1
  its influence on Lorentz 99–100
  diffraction theory 14–15
  double refraction theory 15–17
  partial-dragging (convection) coefficient 100, 101, 242–3, 250, Paper 1
  sine and tangent laws 14, *18*, 54, 58, 89, 176, 185
  theory of polarization by reflection and refraction *18*
Functions of the aether 3–4

Gauss, K. F. 254
Giese, W. 252–3
Goldberg, S. viii, ix
Glazebrook, R. T. viii, 59, 67–9, 84
Gravitation 4, 93, 246
Green, G.
  aether theory of double refraction 58 *see also* W. Thomson
  aether theory of reflection and refraction *46–58*, 69, 71, Paper 4
Green's functions 46

Hamiltonian form of Principle of Least Action 44, 92, 98, 214–7
  derivation from Lagrangian mechanics 43–4, 207
Hamiltonian Principle in Maupertuis' form 44
Hamilton's proof of conical refraction 17
Haughton 59
Heaviside, O. vii, 82, 84n, *89–90*, 95
  on aberration 101
Heaviside's "rotational aether" Paper 8

INDEX

Helmholtz, H. von 68, 76, 95, 100, 252
Hertz, H. 76, 106, 227
Hesse, M. B. ix
Hirosige, T. viii, 99
Holton, G. viii
Huygens, C. *8–11*, 16, 61, 213

Jamin, M. J. 58
Jammer, M. 95n

Kelvin, Lord *see* W. Thomson
Ketteler, 68
Kirchhoff, G. vii, 67
Korn, A. viii
Kuhn, T. S. 5–6

Lagrangian mechanics 41–3, 207, 214
    Green's use of 41, 162, 164, 166
    MacCullagh's use of 41, 190, 192
Larmor, J. 21, 45, 66, 89
Larmor precession
Larmor's aether theory *91–8*, Paper 9
    applied to moving bodies 110–12, 230–46
Larmor's correspondence (correlation) between stationary and moving systems 235–8
Larmor's proof of the intertranslatability of Green's, MacCullagh's and Kelvin's aethers 71
Liènard, A. 105, 268
Light *see* Wave theory of, Corpuscular theory of
Lindsay, R. B. and Margenau, H. 43
Lines of force 77–8
Lloyd, H. viii, 16
Local time 104, 108, 257
Lodge, O. 91n, 208
Lommel 68

Longitudinal waves 14
    in aether, difficulties with 47–8
Lorentz, H. A. 72–3, 145–6, 155
Lorentz aether 68
    as an absolute reference frame 106, 113
    as non-mechanical 106–7
    as possessing substantiality 115
Lorentz
    criticism of MacCullagh's aether 65
    criticism of Michelson interferometer experiment 36–7
    criticism of Planck's ponderable compressible aether 38
    criticism of Stokes' aberration theory 38, 103, 247–8
    dispersion theory 100
Lorentz electron theory 96–7, *99–116*
    derivation of Fresnel convection coefficient 100
    fundamental equations of, in a moving system 108–9, 256–7, 264–5
    fundamental equations of, in a rest system 107, 256
Lorentz–Fitzgerald contraction 91, *103–4*, 109, 111
Lorentz' Theorem of Corresponding States
    to the first order of v/c 110, 267
    to second and higher orders of v/c 110, 112–3, 271
    and Larmor's priority on extension to second order 110–12
    See also Larmor's correspondence between systems
Lorentz' theory of reflection and refraction 84, 100
Lorentz transformation equations 104–5, 108, 257, 269
    See also Lorentz' Theorem of Corresponding States
Lorenz, L. V. 65

MacCullagh, J. vii
MacCullagh's aether theory 59–68, 71, 227, Paper 5
 generalized by Fitzgerald 85–9
 generalized by Larmor to include electrons 217, 219–23
 See also Stokes, Whittaker
Mass
 as electromagnetic in nature 95, 246
 variable with velocity through the aether 105, 272
Maupertuis 213 *see also* Hamiltonian Principle...
Maxwell, J. C. vii, 78–84
Maxwell's electromagnetic theory 3, 79–84, 97, 204–5, 208, 250, 254
 applied to moving bodies 100
 criticized as needing a more specifically characterized aether 84, 84n
 Kelvin's view of 69
 See also Electromagnetic aether
McCormmach, R. viii, ix, 95, 99
Mechanical explanation 5, 60, 107
 see also Reduction
Mechanical model(s) (realizations) 5, 79, 106–7, Paper 6
 See also Analogy
Michelson–Morley interferometer experiment of 1887 vii, 4, 242
 historical background of 32–9
 Lorentz' response to 101, 104, 110, 248, 268, 272
Michelson–Morley moving water experiment of 1886 29, 249
Molyneux, S. 21

Nagel, E. 75
Navier, C. 18, 45
Neumann, F. 59, 65
Newton, I. 8–11

Oersted, H. C. 4

Permeability of aether and matter 126–7, 249, 255
Perpetual motion, denied by Green 165
Photoelectric effect 20
Planck's ponderable compressible aether 7, 38
Plenum, aether as a 7
Poincaré, H. 112–13
Poisson, S. D. 18
Polarization 13, 18, 55, 62–3, 174–180, 186, 191–3
Popper, K. R. 6
Preston, T. 17

Quantum mechanics 99
Quaternion notation 85, 85n, 86–8, 204–7

Rayleigh, Lord *see* Strutt, J.
Rayleigh–Brace experiment 112
Reduction 75
 Larmor's views on dynamical 214
Res Extensa 7
Rosenfeld, L. viii, 76

Saint-Venant, B. de 59
Sarrau 59
Sellmeier, W. 68
Shankland, R. 33
Sommerfeld, A.
 on MacCullagh's and Fitzgerald's aethers 89n, 93n
 on the difficulties of reading Maxwell 82
Space transformation *see* Lorentz transformation equations

INDEX

Stellar parallax *see* Bradley's discovery of stellar aberration
Stokes, G. G. vii, viii
  aberration theory *29–32*, 36–7, 39, 247–8, Paper 2
  criticism of MacCullagh's aether 61, *64–8*
    Larmor's views of 93–4
    outflanked by Kelvin model 71
  diffraction theory 15, 66
  on the plasticity and rigidity of the aether 66–7
Stoney, G. J. 95
Strutt, J. (Lord Rayleigh) 38, 46, 57
  criticism of the MacCullagh–Neumann aether 65
Swenson, L. 33

Thomson, J. J. 76
Thomson, W. (Lord Kelvin) vii, 76, 90n, 95, 97
  mechanical (gyrostatic) models of the aether 60, 66, *68–75*, Paper 6
    Larmor's comments on 94
  modification of Green's aether theory 46, *69–71*, 201, 201n, 209
Time *see* Local time
Time-dilation, Larmor's 241
Transverse waves 14
  in aether 162–3, 189
Tricker, R. A. R. 77, 79
Trouton–Noble experiment 112

"Unified field theory" approach to the aether 3–4
Ur-aether 93–4

Verdet, E. 16–17
Voigt, W. 68
Vortex atom 95, 229
Vortex-sponge aether viii
Vortices
  absence in Stokes' aether 32, 140
  Maxwell's 79

Wave theory of light (undulatory theory) 8, 11–12, 136, 144–5
Weber, W. 76, 253–4
Whittaker, E. T. viii, 76, 79, 116–7
  comments on MacCullagh's aether 59–60, 65
Wien, W. 95
Williams, L. P. 77

Young, T. 135
  on aberration 23
  on his wave theory of light 11–13
  on the nature of the aether 11–12, 55

Zeeman effect 112